Automotive Lightweighting Using Advanced High-Strength Steels

Other SAE books of interest:

Dictionary of Materials and Testing, Second Edition
By Joan Tomsic
(Product Code: R-257)

Mechanics Modeling of Sheet Metal Forming
By Jwo Pan and Sing C. Tang
(Product Code: R-321)

Lightweight Magnesium Technology 2001–2005
By Thomas Ruden
(Product Code: PT-131)

Automotive Lightweighting Using Advanced High-Strength Steels

By Paul Geck

Warrendale, Pennsylvania
USA

400 Commonwealth Drive
Warrendale, PA 15096-0001 USA

E-mail: CustomerService@sae.org
Phone: +1-877-606-7323 (inside USA and Canada)
+1-724-776-4970 (outside USA)
Fax: +1-724-776-0790

SAE Order Number R-431
DOI 10.4271/R-431

Library of Congress Cataloging-in-Publication Data

Geck, Paul, 1982–
Automotive lightweighting using advanced high-strength steels / by Paul Geck.
pages cm
Includes bibliographical references and index.
ISBN 978-0-7680-7978-4
1. Steel, Automobile. 2. Lightweight steel. 3. Automobiles—Materials—Research.
4. Automobiles—Weight. I. Title.
TL240.5.S74G43 2014
629.2'32—dc23
2014006347

ISBN-Print 978-0-7680-7978-4
ISBN-PDF 978-0-7680-8129-9
ISBN-ePub 978-0-7680-8131-2
ISBN-.prc 978-0-7680-8130-5

Table of Contents

Preface

In the early 1990s, as a technical specialist at Ford Motor Company, I had a small support staff working for me to put together and solve large vehicle system computer-aided engineering (CAE) models. Though my work was primarily focused on steel, I also did some investigations on alternative materials (e.g., aluminum and magnesium). It was around this period that Al Gore spearheaded the Program for Next Generation Vehicles (PNGV) during the early years of the Clinton administration. Materials research played a big part in that program, and aluminum came to the forefront as the material of the future for automobiles. Most of the material focus was on the structural body of the vehicle, and the claim was made at that time that 50% of the weight could be saved by converting steel to aluminum; mid-sized concept vehicles were built by General Motors and Ford to prove that claim.

During the early PNGV days, through my own personal research, I came to believe that the claims for aluminum weight reduction were probably being exaggerated, but may not have been that far off, in that cars of that era had somewhat unoptimized architectures and were built out of fairly low-strength steels. The American Iron and Steel Institute (AISI) countered PNGV by building the Ultralight Steel Auto Body (ULSAB), which demonstrated that weight reductions close to what was claimed by PNGV could be achieved by extensive application of high-strength steels and accommodating architectures. This program also introduced advanced high-strength Steels (AHSS), primarily dual-phase steels, which were used to replace some of the previous generation mild steels and conventional high-strength steel (CHSS).

Though Corporate Average Fuel Economy (CAFE) requirements remained the same, during this period, the National Highway Traffic and Safety Administration (NHTSA) along with the Insurance Institute of Highway Safety (IIHS) were busy introducing new safety requirements along with test procedures, which became de facto requirements. This, along with pending fuel economy regulations in the United States and high fuel prices in other parts of the world, prompted a number of material initiatives by the International Iron & Steel Institute (IISI) (e.g., ULSAB-AVC and Future Steel Vehicle) and by PNGV, which morphed into Freedom Car under the Bush administration. Also, by this time, Freedom Car had become more material neutral and supported some amount of steel research. Steel research was also supported by the Auto/Steel Partnership (A/SP) in the United States and internally at the auto and steel companies.

Throughout this period, leading up to my retirement at the end of 2006, many materials engineers throughout the automotive industry still considered aluminum the material of the future for automobiles and still believed the weight reduction potential for aluminum to be about 45%, despite all of the steel research to the contrary. For this reason, I started to teach a seminar at the Society of Automotive Engineers on AHSS that would lay out the true technical case for how much weight could really be taken out of automobiles, using the different material alternatives. However, I believe that the course I'm teaching has hardly moved the needle for both the alternative materials engineer and the steel engineers. As evidence of this, AHSS account for less than 10% of the steel sheet metal volume used prior to the time of this writing (2013). However, the automobile industry is faced with the greatest hurdle, in terms of regulation, that it has ever faced. The Environmental Protection Agency (EPA) is now requiring that automobiles achieve a fleetwide average of 54.5 MPG by 2025 through a very aggressive fuel economy phase-in program. To meet this requirement, the auto industry is taking a very hard look at alternative materials, primarily aluminum, and is accelerating the application of AHSS. Meanwhile, the steel industry is starting to market second and third generation AHHS.

In support of the latest automotive challenges in terms of weight reduction, I wrote this book in an attempt to lay out the true opportunities for alternative material utility in automobiles and to offer the most up-to-date design guidance in efficient architectures supported with the application of AHSS while exploring weight savings and resulting fuel economy advantages of this strategy. Realistic comparisons with other alternative materials are made through detailed analysis. Many projects that I have been part of will be explored to demonstrate how AHSS technology has developed and to get us to the foothills of the mountain that we, as automobile design people, now must climb.

Paul Geck

Acknowledgments

This was a difficult book to write in that it covers several disciplines (e.g., automotive design engineering, computer-aided engineering, vehicle attribute engineering, metallurgy, automotive manufacturing engineering, book editing, and so on). I have strengths in some of these areas but would not be considered an expert in many, especially not in metallurgy or in book editing. Therefore, I had to draw on the strengths of several colleagues. First, I would have to recognize Martha Swiss and Heather Slater of the SAE for their encouragement and editing skills. Then I would have to thank Rich Cover, who is my co-instructor for the SAE's Advanced High-Strength Steels for Vehicle Weight Reduction class and who helped fill my metallurgy and steel-making knowledge gaps. I would also like to thank Ron Krupitzer, the vice president of automotive applications at the American Iron & Steel Institute, who helped pick a couple of steel-automotive experts to review my manuscript as it was being developed and guided me through some of the more difficult passages. Those experts were Blake Zuidema, ArcelorMittal's director of product applications, and Dean Kanelos, automotive development metallurgist at Nucor Steel. In terms of contributions to specific sections of this book, I drew significant portions of Chapters 4 and 5 from Harjinder Singh's work as the project manager for the National Highway and Traffic Safety Administration/EDAG Inc. lightweight vehicle (LWV) study and the NanoSteel project, which was an outgrowth of the LWV study. I also drew material from Jody Shaw's 2013 Great Designs in Steel talk to support Chapter 4. Jody is the director of technical marketing and product research at United States Steel Corporation. I received support for many of the figures I used from Deanna Lorincz, who is the senior director of communications for the Steel Market Development Institute. Finally, I have to thank the personnel of the NanoSteel Company (including Ellen Bossert, chief marketing officer) for supplying much of the material for Chapter 5, which points to the future of advanced high-strength steels.

Introduction

I.1 Scope

This book is written with the original equipment manufacturer (OEM), Tier 1, and Tier 2 automotive design engineers in mind. However, it is hoped that many of the support people (e.g., metallurgists in both automotive and steel companies, test engineers, computer-aided engineering [CAE] people, materials purchasing personnel, product planners, and management) will also find this information useful. The focus will be on body and chassis structures and the sheet metal, of which these systems are primarily comprised of. More of the material addresses the automotive body, as this is where most of the advanced high-strength steels (AHSS) are being applied today. The fundamental purpose of this book is to provide information that will be useful when engineers embark on the designs of their next generation of vehicles, so that they can make informed decisions on what basic materials to use and how to optimize those materials to achieve cost-effective weight reduction. Of course, in this book the emphasis will be on steels in general and AHSS in particular. However, there is much information on comparisons of steel with alternative materials for different subsystems of the vehicle.

The past, present, and future of advanced high-strength steels (AHSS) will be covered in this text as well as competing technologies such as aluminum sheet metal. Most of the material is concerned with North American applications of AHSS, because this is more familiar territory for the author, though applications of AHSS in Japan and Europe actually predated North American (NA) applications, driven by the higher fuel costs there. Given that, I have included some information on AHSS applications and studies from other parts of the world. For this book, advanced high-strength steels are defined as follows:

> AHSS is a family of steels having higher strength than most steels but with better formability than today's conventional high-strength steels. AHSS are typically multiphase steels with some percentage of martensite (the strongest of the steel phases). For our purposes, the martensitic and the hot forming steels will also be considered AHSS. For the most part, we will consider all ultra-high-strength steel as being a subset of the AHSS.

I.2 History of Advanced High-Strength Steel

At present, most automobiles and light trucks are approximately 70% ferrous by weight for the whole vehicle, depending on segment. The percent of alternative material utilization has leveled off over the last few years, as it has become increasingly difficult to find new, cost-effective, alternative material applications. At the same time, new U.S. safety and fuel economy regulations have intensified pressures at all the OEMs to seek more aggressive measures for weight reduction. These pressures are causing auto companies to rethink alternative material applications and to look for remaining steel opportunities.

In the early-1980s, the major automotive OEMs were on par in terms of weight reduction/avoidance technologies. In the mid-1980s, some of the auto companies started to introduce high-strength steels (e.g., high-strength low-alloy [HSLA] and bake hardenable [BH] steels). For example, the 1986 Ford Taurus had approximately 80% mild steel (i.e., below 180 megapascals [MPa] yield strength) and 20% high-strength steels. The high-strength steels, which were used on the Taurus, were between 180 and 350 MPa yield strength, which we now call conventional high-strength steel (CHSS). Some believed by the early 1990s that cars were at the limit of high-strength steel utilization, at least from a reduced thickness perspective, because further thickness reduction would bring major vehicle stiffness degradation and might be too difficult to form or weld. However, during the 1990s many auto companies, primarily Japanese and European, were able to surpass the 20% high-strength utilization and to progress into the realm of AHSS. Toyota and BMW were believed to be leaders in the application of new steel technology. Higher fuel prices in Japan and Europe provided the impetus for more concentration on weight reduction through stronger steels.

Because the availability of the new steels increased in North America and because of the looming new safety and fuel economy requirements, the NA car companies started to introduce AHSS in the early 2000s. An example of this was the General Motors Epsilon platform as an early leader in AHSS. In this decade, with the advent of the new regulations, an even more aggressive AHSS emphasis is occurring. However, this is accompanied with a similarly aggressive look at alternative materials to steel.

I.3 Steel Technology Studies

In the early 1990s, the Ultralight Steel Auto Body (ULSAB) Consortium of 35 steel companies from around the world commissioned Porsche Engineering Services (PES) to design a lightweight steel D-class body-in-white partly in response to the emphasis on aluminum within Al Gore's Program for a New Generation of Vehicles (PNGV). The ULSAB project was completed in 1998, and PES developed a design and built a body demonstrator, which was between 25% and 45% lighter, depending on which of the many benchmarked vehicles one compared the results to. Though this study was somewhat unconstrained, it is interesting in retrospect that many of the technologies have found their way to at least one product in the world. This can be seen in the increased usage of AHSS, hydroform technology, and laser welding. However, not all of the technologies have become standard practice, even though most of the technologies were probably directionally correct.

The original ULSAB study was then followed by the ULSAB-AVC (advanced concept vehicle) study in which C-Class and PNGV class vehicles were designed. The AVC study was a total vehicle study and was even more aggressive than the initial ULSAB study in terms of architectural features and technology stretch.

An example of a similar study, which had auto industry participation (Ford) and was completed in 2001, was the Improved Materials & Powertrain Concepts for 21st Century Trucks (IMPACT) program, which was partially funded by the U.S. army and the North American Steel Suppliers. This program also had the University of Louisville and Mississippi State as major participants. One of the goals of this project was to take 25% of the weight out of the then current Ford F150 in the most affordable way possible. The IMPACT study did reach its goal of 25% at the vehicle level, while achieving 20% reduction in the body-in-white weight. Also, the goal was reached while only incurring about a $350 variable cost penalty. More interesting was the fact that 21% vehicle weight reduction could have been reached at no variable cost increase, and a 12.5% weight reduction could have been achieved at a $450 variable cost reduction. Much of the cost-reducing weight reduction was achieved through a shift to the higher-strength steels. This study was also constrained in that the redesigned vehicle met the baseline F150 vehicle attribute levels.

The ULSAB-AVC and IMPACT studies were followed by the Future Steels Vehicle (FSV) Study, which was sponsored by WorldAutoSteel, a consortium of major steel companies worldwide. This study was more futuristic than the other studies in that a major shift in vehicle architecture was assumed to accommodate new powertrain architectures and a new regulatory environment.

More recently, two studies were commissioned by the Environmental Protection Agency (EPA) and the National Highway Traffic and Safety Administration (NHTSA), respectively, to investigate cost-effective weight reduction in a particular vehicle. In the case of the EPA study, the vehicle was the 2010 Toyota Venza cross-over utility vehicle (CUV), and in the NHTSA study the vehicle was the 2010 Honda Accord. EDAG Inc. was the primary contractor for the NHTSA study and was one of the contractors for the EPA study. The NHTSA study is probably more relevant for this book in that aluminum and steel are compared for major subsystems for the same vehicle. That makes the NHTSA study somewhat unique for material studies, which are in the public domain. The only shortcoming of the NHTSA study was that it focused on first generation AHSS for its steel solution. In this book, we will try to extend the NHTSA/EDAG study to outcomes inclusive of second and third generation AHSS.

Chapter 1
Advanced High-Strength Steel Technology

1.1 Characteristics and Metallurgy

This chapter will serve to define advanced high-strength steels (AHSS); to discuss the differences between AHSS, mild steels, and conventional high-strength steels (CHSS); and to review the ways of making AHSS. First, it will be instructive to discuss the characteristics and metallurgy of AHSS. To characterize steels and other solid metals, a stress-strain curve is produced from a machined bar or strip sample of the material (Figure 1.1).

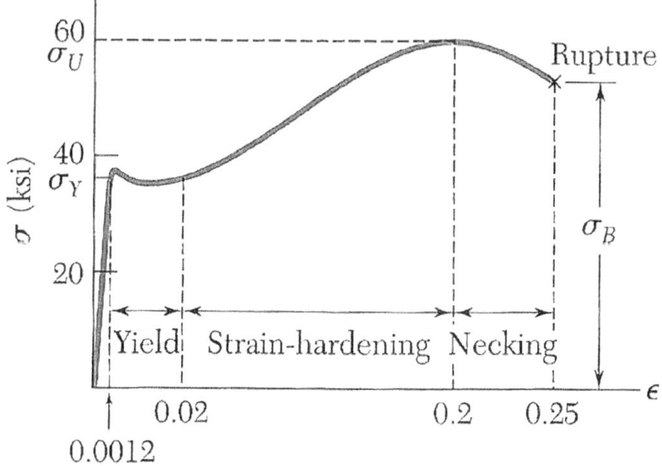

Figure 1.1 Stress-strain curve

The test specimen is put in a tensile test machine and pulled until it breaks. Figure 1.1 shows the engineering stress versus the engineering strain for low carbon steel. The engineering stress is defined as the force, pulling the sample, divided by the original cross-sectional area of the sample. The strain is defined as the extension of the sample divided by the original working length of the sample. The stress is in pressure units (i.e., force per square unit of area). For this book we will be using SI units for the most part, which means that we will be using megapascals (MPa) for stress. One thousand pounds per square inch (KSI) equals 6.895 MPa. Strain is expressed as a percent because it is a length dimension divided by a length dimension. Often, the stress-strain curve is presented in true stress versus true strain as opposed to engineering stress versus engineering strain. True stress is defined as the force on the sample divided by the instantaneous cross-sectional area at the center of the sample; true strain is defined as the change in length divided by the instantaneous length at the time of measurement. The true stress-strain curve is similar to the engineering stress-stain curve at low strain levels, but it typically does not decrease after the ultimate engineering tensile stress is reached (necking). For steels in general, the stress is proportional to strain for low strain levels (i.e., the elastic region), until the curve reaches the yield stress. The steel exhibits plastic behavior until the ultimate stress is reached, and then the engineering stress decreases due to localized necking until the sample breaks. The strain at which breakage occurs is the total elongation. Another useful property of the stress-strain curve is the n-value or work hardening exponent. The n-value is derived from the slope of the true stress versus true strain plot in the plastic region over a defined strain range. Typically, a steel with a higher n-value is more formable than steels that otherwise have equivalent strength characteristics. Another metric, which is calculated from the stress-strain test, is the r-value. The r-value is a measure of the resistance to thinning and is calculated from the change in thickness of the stress-strain test specimen. A higher r-value means that the specimen will experience less thinning during the test. The r-value (as with the n-value) is a measure of formability.

Another way of characterizing steels is by their microstructures and relative strength levels, as shown on the banana curve (Figure 1.2).

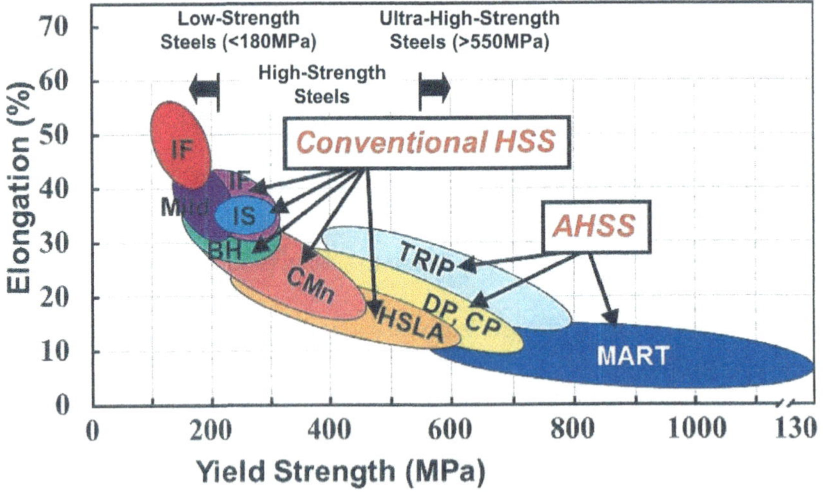

Figure 1.2 Banana curve

In this graph, elongation is plotted versus yield strength. Different classes of steel fall within the colored ovals on the graph. For instance, interstitial free (IF) steels are fairly ductile, low-strength steels. Within a class of steels, there might be several different specific strengths or grades available. IF 140 and IF 160 are two different grades of IF steels. Low-strength steels typically are steels that have less than 200 MPa yield strength and have greater than 30% elongation. Some of the low-strength steels are the mild steels and IF steels. Some IF steels (such as IF-rephos, rephosphorized interstitial free steels) that have greater than 200 MPa yield strength are typically classified as CHSS. Other types of CHSS are isotropic (IS), bake hardenable (BH), carbon manganese (CMn), and high-strength low alloy (HSLA). The AHSS have typically more strength than the CHSS and/or have more ductility than the CHSS at the same strength level. Dual phase (DP), complex phase (CP), transformation-induced plasticity (TRIP), and the martensitic (Mart) are all considered AHSS.

To understand how AHSS are made and what is physically different with AHSS, it is useful to study the microstructure of AHSS (Figure 1.3).

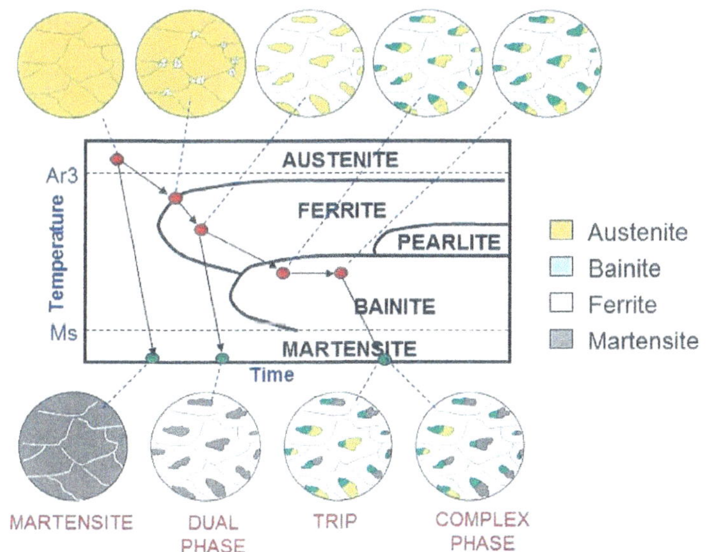

Figure 1.3 Microstructure of AHSS

The rectangle at the center of Figure 1.3 is a time-temperature-transformation (TTT) diagram for a carbon steel. The alloys used to produce steels will change the shape and location of the lobes on this chart. Every unique chemistry has its own unique TTT curve. The dot at the upper left-hand corner of this diagram represents the starting point for a steel, which has been heated to or above its austenitic temperature (about 700 °C or 1300 °F). Steels are annealed (i.e., heated into the austenitic range) after cold rolling to get back to the original properties. How fast the specimen is cooled will determine what the microstructure will look like. The lobes on the diagram determine what phases (i.e., austenite, ferrite, pearlite, bainite, or martensite) will exist in the specimen after it is cooled down. If it cools very slowly, it will have a ferritic structure, which is a very ductile but low-strength phase. Most steels (mild steels) are ferritic. If it is cooled very rapidly, the austenite will convert directly to martensite, which is the strongest, but most brittle phase. Depending on the cooling rate, the microstructure, in general, will be a combination of phases. For instance, dual-phase steel, which is a combination of ferrite and martensite, is the most common of the AHSS. It has a strength and ductility somewhere between ferrite and martensite. Though austenite is an unstable, ductile phase, which converts to other phases at room temperature, alloying can produce steel, which has retained austenite in it. TRIP has some retained austenite in it, which converts to martensite on work hardening. This gives TRIP a little bit more ductility and a higher work hardening rate than a DP grade with the same strength. Complex phase is an AHSS, which is a combination of several phases.

Besides the common types of AHSS, there are new generations of AHSS which haven't been used in automotive structural applications yet, but which may be used in the future as more and more weight reduction is necessary to meet fuel economy targets. These are the second and third generation AHSS, shown in the added balloons to the banana chart in Figure 1.4.

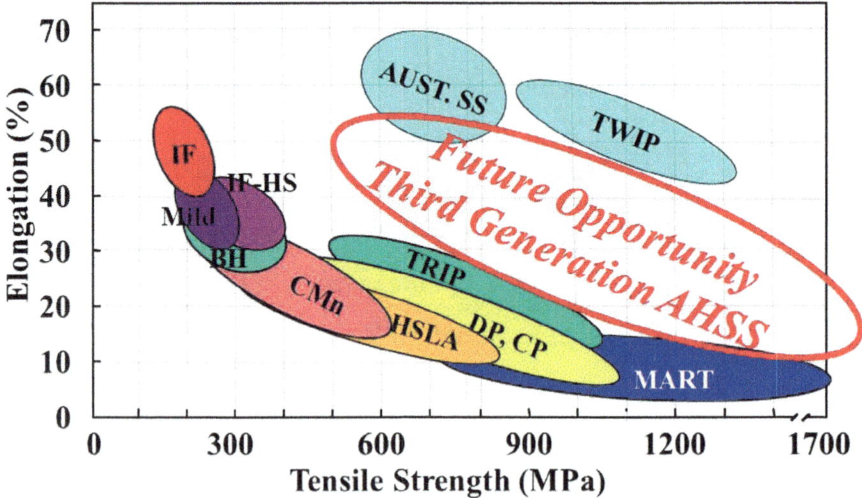

Figure 1.4 Generalized banana curve

Twinning-induced-plasticity (TWIP) and austenitic stainless steels are considered second generation AHSS and provide as much or more ductility than mild steels but at the strength levels of ultra-high-strength AHSS.

These steels have a great deal of alloying, which gives rise to relatively higher costs, and this is why they haven't yet broken into automotive structural use. However, austenitic stainless is used commonly in other, less cost-sensitive applications. Of more concern to us are the third generation AHSS, which provide AHSS strength and much improved ductility and at expected costs somewhere between the first and second generation AHSS. These steels are also not in production for automotive use so far, but the steel industry is working to create grades that will fall into this category. The steel producers hope that steels with these properties, and costs at roughly half that of aluminum, will offset the temptation to convert to alternative, nonsteel materials. In this book we will primarily address the first generation AHSS, as these steels have not been fully exploited in the marketplace, but later we will try to quantify the added benefits of the third generation AHSS.

To get a better idea of the properties of AHSS, it is useful to compare these steels with CHSS, which have somewhat similar strength levels. In Figure 1.5, a DP steel and an HSLA CHSS with the same yield strength are compared.

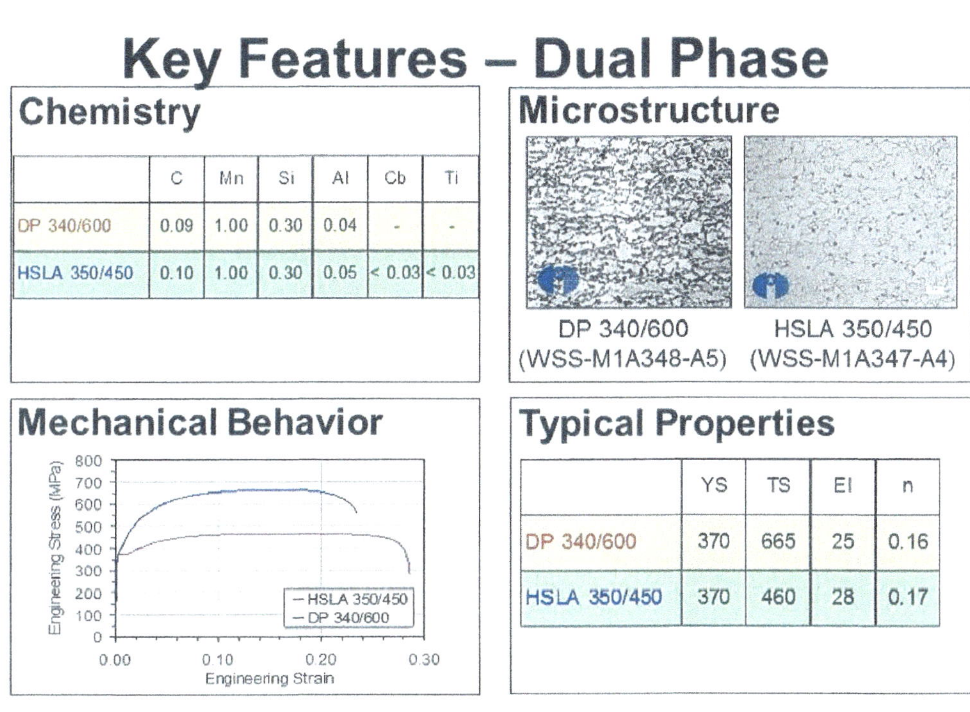

Figure 1.5 Dual-phase versus HSLA steel

In order to avoid confusion on grade designations, it is useful at this point to mention that if a steel grade is designated as DP 340/600, the first number refers to its yield strength and the second number refers to its ultimate tensile strength ("tensile strength"). However, for AHSS it is common to refer to the steel grade only in terms of the tensile strength (i.e., DP 600). However, most steel manufacturers are producing DP 600 as having a minimum tensile strength of 590 MPa versus a nominal 600 MPA. Therefore, when one talks of DP 590 or DP 600, they are really referring to the same grade of steel. For CHSS, if a single number is used to refer to it, the number is typically the yield strength. HSLA 350/450 is often designated as HSLA 350, where 350 MPa is the minimum yield strength. The primary reason the AHSS grades are designated by tensile strength is because they were first adopted by the Japanese auto companies, whose grades are all designated by their tensile strengths. Though the use of two different methods for specifying AHSS and CHSS is often confusing, it is helpful to realize that CHSS are many times selected based on their yield strength, whereas AHSS are mostly selected based on their tensile strength.

In Figure 1.5, the microstructures shown in the upper right-hand corner are what these grades of steel would actually look like under a microscope. The chemistry or the added alloying elements are fairly similar for the two steels. The typical yield strength (i.e., 370 MPA) is the same, not to be confused with the minimum yield strength or the nominal yield strength. The elongation and the n-value are only slightly higher for the HSLA steel. The big difference is in the much higher tensile strength provided by the DP steel, which is due to a modified chemistry and a different cooling rate when the steels are annealed. Since the yield strength, elongation, and n-values are close for the two steels, one would expect that formability would be similar for the two steels. Another comparison, which might be useful, is TRIP 600 compared to HSLA 350 (Figure 1.6).

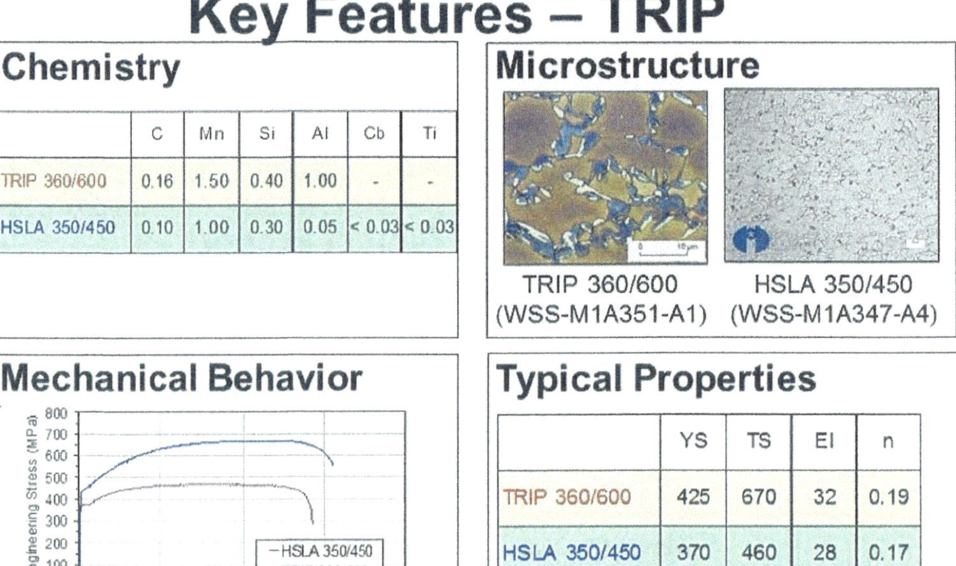

	C	Mn	Si	Al	Cb	Ti
TRIP 360/600	0.16	1.50	0.40	1.00	-	-
HSLA 350/450	0.10	1.00	0.30	0.05	< 0.03	< 0.03

	YS	TS	El	n
TRIP 360/600	425	670	32	0.19
HSLA 350/450	370	460	28	0.17

Figure 1.6 TRIP versus HSLA

This TRIP comparison is being made because TRIP 600 has a yield strength similar to HSLA 350. Note that in this case the elongation and n-value are higher than the HSLA 350. This would indicate that TRIP 600 formability should be better than both the HSLA 350 and the DP 600. However, the yield strength is higher for the TRIP, which might indicate more springback on forming. We will discuss springback in the next chapter. Also, notice that the chemistry is richer for the TRIP. So TRIP can be expected to be more expensive than the HSLA and DP steels. Finally, we will compare martensitic steel to HSLA (Figure 1.7):

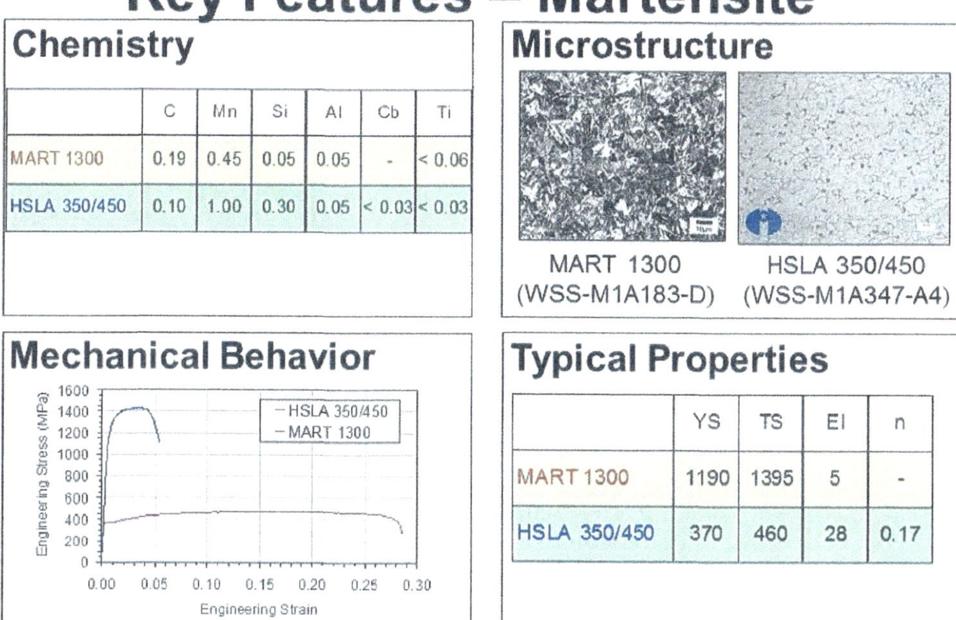

Figure 1.7 Martensite versus HSLA

You can see that martensite, aside from high carbon levels, is not so rich in chemistry. The most notable feature of martensite is its high strength levels, both yield and tensile. Of course the high yield strength, low elongation, and nonexistent n-value mean that it is very difficult to form. This is why martensitic steels are usually roll-formed (which we will discuss later). A variant of martensitic steels are the hot-formed steels. These steels are alternately called boron steels, because boron is one of the additives, or press-hardened steels (PHS). These steels are formed hot and typically quenched in the die. Though these steel grades are fairly expensive when considering the extra processing cost, their high strength and their ability to be heavily formed with good dimensional tolerance make these the fastest growing steel grades in the AHSS lineup.

1.2 Drivers for Advanced High-Strength Steels

What is driving auto manufacturers to increase their utilization of AHSS? Safety is the most direct driver for AHSS. If you look at Figure 1.8, you can see a steady progression of new National Highway & Traffic Safety Administration (NHTSA) required safety regulations or de facto regulations going back from the early 1990s through the present time. De facto regulations are test procedures and metrics that are proposed by the Insurance Institute of Highway Safety (IIHS). Though compliance with these metrics is not required by law, the political pressure put on the auto manufacturers by the IIHS means that compliance is for the most part mandatory.

Safety Standards Growing Increasingly More Stringent

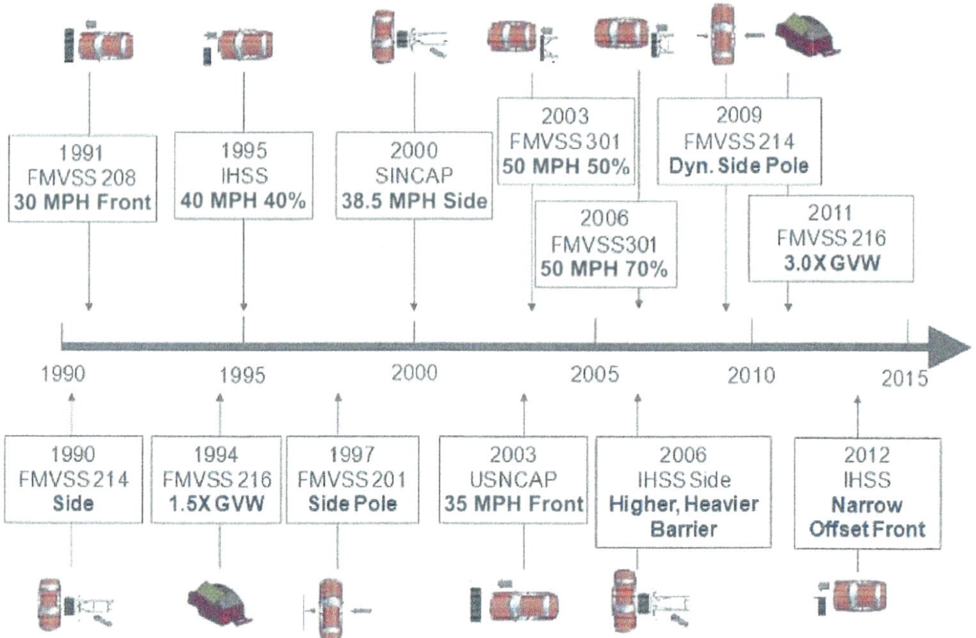

Figure 1.8 Safety requirements

Thirty years of rapidly escalating safety requirements have been the primary cause of increased vehicle weight over this time period. In the late 1980s and early 1990s, noise, vibration, and harshness (NVH) were the leading vehicle attributes contributing to weight increases. Today, if a vehicle satisfies the safety requirements, it will probably be acceptable for NVH as well as durability.

For the NHTSA United States New Car Assessment Program (US NCAP), 35 MPH (58 KPH) full-frontal impact, the pass-fail criteria involves minimizing the dummy head impact criteria (HIC) and the deceleration of the dummy's chest, which are usually expressed in Gs. For the vehicle to perform well for this criterion, longitudinal forces need to build up gradually but reach a very high force level at the end of the crash event (Figure 1.9). This will ensure that the maximum amount of energy is absorbed. Having fairly low yield strengths but very high tensile strengths makes the DP and TRIP grades ideal for this process. Most manufacturers are using DP 800 for their newer models in the front and rear crush components (e.g., rails), as ultra-high-strength AHSS may produce a collision that is too abrupt and would tend to injure the driver and passengers. However, bumper components require very high strength martensitic steels to meet damageability requirements.

The significance of using DP steels, which tend to absorb more energy than CHSS, can be seen in Figure 1.10, where 2.00 mm DP 600 is compared to 2.5 mm HSLA 350 for the front rails.

Crashworthiness Fundamentals

Engineering Stress (MPa)
700
600
500
400
300
200
100
0

0.0 0.1 0.2 0.3 0.4 0.5
Engineering Strain

Capacity to absorb crash energy is indicated by energy under stress-strain curve at 10% strain and 100 s^{-1} strain rate

Energy absorbers (front, rear long. components) benefit from DP/TRIP grades up to 800 MPa UTS:

- **Large area under σ-ε curve for energy absorption**

Figure 1.9 Crashworthiness fundamentals

Safety/Crash Energy Management

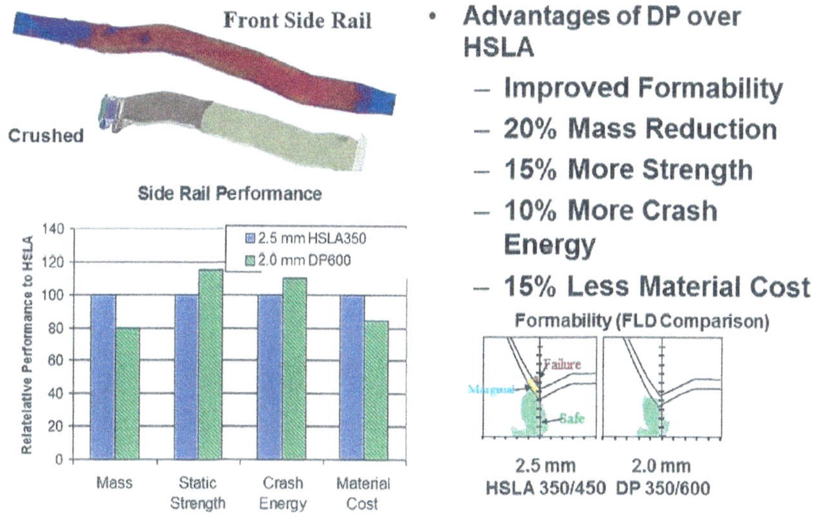

Figure 1.10 Safety/crash energy management

The front rails control much of the energy management for the front end. You can see that the DP 600 provides 20% mass reduction, 15% more strength, and 10% more energy absorption; and it does this at 15% less material cost. Even though the DP is more expensive on a weight basis than HSLA, there is a lot less of it, having reduced the thickness by 20%. The DP is also more formable than the HSLA, as shown in the lower right-hand corner of Figure 1.10, where the forming limit diagrams are compared. We will be discussing forming limit diagrams in more detail later in this book.

However, most of the safety requirements are oriented to provide an intact safety cage around the driver and passengers. This means that auto manufacturers for the most part are trying to use the strongest steels, both yield and tensile strengths, for components, which protect the safety cage. Of course, formability is one limiting factor of how much strength can be used in the various body components.

As you might imagine, weight reduction is equally responsible for the increased use of AHSS because it directly affects fuel economy. The Environmental Protection Agency (EPA) has set a goal of achieving an average of 54.5 MPG (14.4 km/l) by 2025. This seemingly impossible goal is causing auto manufacturers to aggressively look at all options in their quest, and of course one of the areas of opportunity is new materials. A joint study undertaken by the U.S. army, Ford, the American Iron and Steel Institute (AISI), the University of Louisville, and Mississippi State University was the Improved Materials and Powertrain Architectures for 21st Century Trucks (IMPACT) Project [1-1]. The goal of this project was to reduce weight for an F150 vehicle by 25%. In this study (Figure 1.11), it was determined that for the F150 a 10% weight reduction would translate to over a 5% improvement in fuel economy.

Weight/Fuel Economy

- **Weight reduction = fuel economy improvement**
- **10% weight reduction > 5% improvement in FE**
- **IMPACT Study: 56% - Over 500 MPa 25% mass savings**

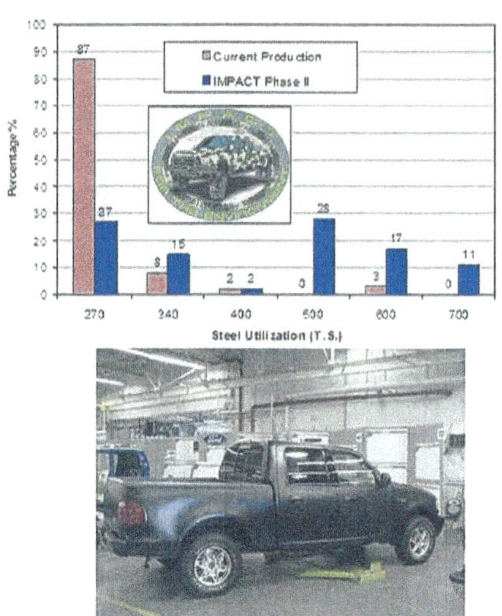

Total Weight Savings - 1310 lbs. (25%)

Figure 1.11 Weight reduction versus fuel economy

The weight savings were achieved in this project primarily by increasing the tensile strength of the steels in the vehicle and decreasing the thickness of the panels while achieving equivalent function.

Of course, one of the primary drivers for increasing fuel economy is the reduction of greenhouse gas emissions (e.g., CO_2). Reducing greenhouse gas emissions using AHSS has been shown to be a better strategy than using alternative materials. For instance, it was shown [1-2] that the payback period in terms of the number of years to offset the greenhouse gas emissions from the production phase is shorter for high-strength steels versus using aluminum. This is primarily because it takes more than 4.5 times the energy to produce a pound of aluminum versus a pound of steel. Also, AHSS are more recyclable than alternative materials:

1. AHSS, like all steel, is 100% recyclable.
2. Steel remains the most recycled material.
3. Steel recycling has a large infrastructure:
    a. 12,000 dismantlers in the United States
    b. 300 shredders
4. Steel autos enjoy a 97% end-of-life recycling rate.

1.3 Growth in Advanced High-Strength Steel Usage

Figure 1.12 shows a projection that was made by one of the steel companies in the mid-2000s, on AHSS usage at five different U.S. auto manufacturers through 2013:

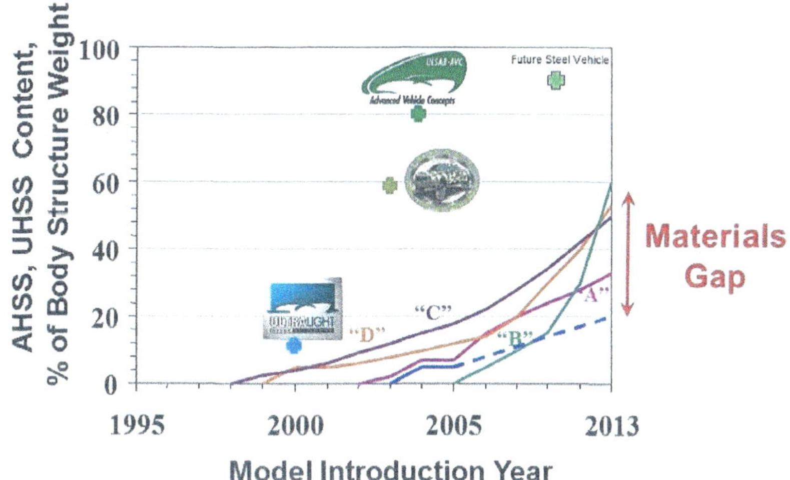

Figure 1.12 AHSS usage in the United States

Superimposed on this graph is the percent of AHSS usage in four benchmark studies that were sponsored at least in part by the steel manufacturers during the same period: the original Ultralight Steel Auto Body (ULSAB) [1-3], the IMPACT, the ULSAB-AVC [1-4], and the Future Steel Vehicle [1-5] projects. There seem to be some differences among the auto manufacturers, but all the manufacturers that were queried were thought to end up between 20 and 60% AHSS by 2013. However, all the manufacturers are significantly below the ULSAB-AVC benchmark of 80% AHSS and the FSV benchmark of 97%.

Figure 1.13 shows the consequences of gaps in AHSS usage in terms of missed weight reduction opportunities.

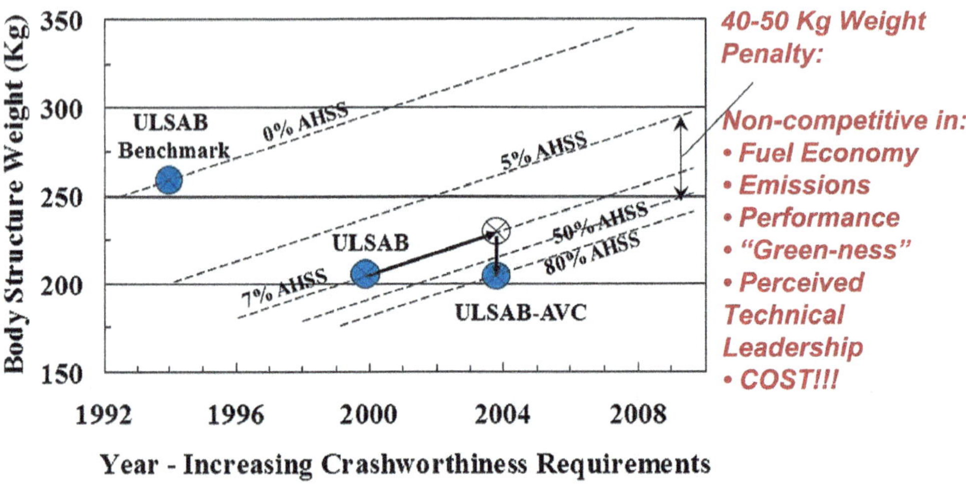

Figure 1.13 Weight penalty for deficiency in AHSS usage

This graph shows that insufficient usage of AHSS can translate into more body structure weight. In particular, 45% less AHSS utilization can translate into 40 to 50 kg (88 to 110 lb) greater body structure weight.

Figures 1.14, 1.15, and 1.16 show the progression of AHSS to the various vehicle parts to 2012 and the recommended applications in the 2017 time frame. Figure 1.14 shows that most manufacturers started out with CHSS (e.g., BH and HSLA grades) in the early 2000s and have progressed to present-time higher DP grades. It is anticipated that the major front-end components will progress to the TRIP and hot-stamped boron (HSB) grades by 2017. Probably the fastest growth area at the present time in AHSS usage is in the HSB area. This is because increasingly greater strength is needed for safety considerations, and because with HSB, formability issues are automatically overcome. However, added processing facilities are required to accommodate the HSB applications. The major facilities required are an oven for heating the parts to very high temperatures, a sophisticated draw die in which cooling ports are drilled for cooling while the part is in the press, and a sand peening facility to get rid of the scale, which accumulates on the part during cooling. The biggest problem is that since the part has to be cooled in the press, the cycle time goes up considerably. This often means that extra press lines have to be built to accommodate high-volume parts. However, the extra facilities and attendant costs are somewhat reduced, for with hot-stamped parts fewer die operations are required (e.g., restrikes) because the part can typically be brought home in the draw die operation. Also, aluminized coated HSB can eliminate the slag and therefore the need for a sand peening operation. However, the base material cost will go up.

Recommended Front/Rear Structure Applications

Primary Consideration: Energy Absorption

	BH	HSLA	DP				TRIP		Martensite			HSB	
Component	300	350	500	600	800	1000	600	800	900	1100	1300	1500	1500
Front upper rails		2003		2008	2012			2012					
Front lower rails		2003		2008	2012			2012					
Rear rails		2003		2008	2012			2012					
Front frame rails		2003		2008	2012			2012					
Rail reinf.		2003/8			2012								
Shock towers		2003/8/10											
Wheel house inr	2008 BH250						2012						

Figure 1.14 Front and rear structure applications

Figure 1.15 shows the progression of passenger compartment AHSS applications.

Recommended Passenger Compartment Structure Applications

Primary Consideration: Minimize Intrusion

	BH	HSLA	DP				TRIP		Martensite				HSB
Component	300	350	500	600	800	1000	600	800	900	1100	1300	1500	1500
Front cross mbr.		2003 HSLA300		2008	2010								
Center cross mbr.		2003 HSLA300		2008							2010		
Rear cross mbr.		2003 HSLA300		2008							2010		
Rocker outer	2008	2003 HSLA300			2010								
Rocker inner		2003 HSLA300		2008	2010								
Rocker reinf.		2003									2008	2010	2010
Dash	2008			2010									
Cowl/Plenum	2008			2010									
Pillar inners				2008	2010			2010					2010
Pillar outers	2008/10												
Pillar reinf.		2003		2008		2010			2010				2010
Roof rail		2003 HSLA300		2008	2010								
Roof bow		2003			2008	2010							2010
Header	2008/10	2003 HSLA300											
Header inner	2008/10	2003											
Package tray	2008			2010									

Figure 1.15 Passenger compartment applications

When compared with Figure 1.14, Figure 1.15 shows a progression to higher-strength steels. This is because passenger compartment components are driven by a need for high strength, both yield strength and tensile strength, whereas front and rear end components are driven by a need for high-energy absorption, where a high tensile to yield ratio is required. Therefore the DP grades will transition to martensitic grades in the 2017 time frame for the passenger compartment components and the heavily formed components will transition to HSB. Some of the formability problems will be overcome with the transition to roll forming versus stamping. However, that will typically require some redesign.

Figure 1.16 shows the chassis applications.

Recommended Frame/Chassis/Suspension Applications

Primary Considerations: Weight, Hole Expansion, Crash

	BH	HSLA	SF/DP/MP				TRIP		Martensite				HSB
Component	300	350	500	600	800	1000	600	800	900	1100	1300	1500	1500
Engine cradle	2008			2010			2012						
Cradle reinf.		2003/8/10											
Suspension arms		2003		2008	2012								
Axle housing		2003		2008/10	2012			2012					
Radiator support	2008												
Frame rails		2003		2008	2012			2012					
Frame cross mbr.		2003		2008/10									
Frame reinf.		2003		2008/10									

*TRIP grades for the more difficult forming applications

Figure 1.16 Frame/chassis/suspension applications

There are some major differences between chassis applications and body applications. First of all, chassis components tend to require thicker gage parts than body applications. Most chassis parts are over 2 mm, whereas most body parts require less than 2 mm sheet metal. This means that most chassis parts are hot rolled, and body parts require cold rolling reduction, which adds to the cost of body parts. This brings up the number one impediment to chassis AHSS applications: If the utilization of AHSS means that a part is reduced below the thickness, which is available in a hot rolled steel, then the auto manufacturer will see a double whammy in terms of cost. The price of the steel will go up because it is AHSS, and it will go up some more because they now have to purchase cold rolled steel instead of hot rolled. Another concern is corrosion. It was generally felt that steels over 2 mm thick could survive without being galvanized but may need to be coated if thinner steels are used.

There are also some differences in available hot rolled grades. For the lower level AHSS, besides the DP grades, there are stretch flange (SF) and multiphase (MP) grades. These grades offer some advantages in formability over CHSS, depending on the type of deformation encountered during the forming operations.

Another impediment to AHSS conversion is that most frames and subframes are stiffness driven (i.e., dependent on elastic modulus) versus strength driven. Since all steels have close to the same modulus, AHSS offers little advantage for those parts. However, all chassis parts require high strength, even those parts where stiffness is the primary requirement. Also, impact strength requirements may be significant for suspension arms. As might be suspected, the transition to AHSS is going slower for chassis driven parts, but it is progressing. One of the advantages that chassis parts have to assist in the conversion is that most chassis parts are made of straight or nearly straight sections. This means that roll forming can be used in many cases, which enables the conversion to martensitic grades.

1.4 Making Advanced High-Strength Steels

It is helpful at this point to discuss steelmaking in general and differences in the production process that need to be made for AHSS production. First of all, the steelmaking process requires a huge installation, which may cover several square miles (Figure 1.17).

Figure 1.17 Steel production facilities

The output of this process is typically steel coils of the type shown in Figure 1.18.

Figure 1.18 Steel coils

These coils are very heavy, up to 25 metric tons (MT). Steel production is often measured in metric tons, which are about 2204.6 pounds, and typically 1500 body parts can be made from a single coil. Each coil makes many of one part, and about .8 MT is required for one vehicle. A typical integrated steel plant can produce about 10,000 MT per day.

Because of physical constraints, the minimum number of coils of a specific grade of steel that a plant will make in a single cast sequence is about 20 coils. However, more typically a plant will make between 60 and 140 coils of a particular grade back-to-back. If the number of coils produced back-to-back is reduced, that will add cost to the process, which is one of the reasons that AHSS costs more.

The core elements of an integrated steel mill are shown in Figure 1.19 [1-7]. An integrated mill contains a blast furnace, which is the conventional method for reducing iron ore to molten iron. Some mills do not use a blast furnace, but use an electric arc furnace (EAF) to melt scrap pig iron and direct reduced iron (mini-mills). Mini-mills sprang up in part due to the high cost of building an integrated mill. Usually EAFs are used to produce long products (structural steel, rod and bar, wire, and fasteners), but some are used to produce sheet. Although there is significantly less steelmaking production using EAFs with thin slab casters, this method still accounts for about 10 million tons of annual U.S. domestic steel production. Mini-mills can produce most of the products made by integrated mills. However, most sheets are produced at integrated mills with the combination of a blast furnace in conjunction with a basic oxygen furnace (BOF). Coal is added to ovens where coke is produced, and iron ore and coke are heated in the blast furnace to create molten iron. However, recently some companies have started to use natural gas instead of coal because the price of natural gas, relative to coal, is coming down [1-6]. The ingredients of the blast furnace are transported to the BOF, where oxygen is introduced to reduce the carbon content of the molten iron to create steel. The BOF is an exothermic furnace where external heat does not need to be added because the addition of oxygen in itself creates heat by reason of the chemical reaction with carbon. In fact, the BOF needs to be cooled by adding scrap to keep the process in control. The steel from the BOF is poured into a ladle, alloyed, and stirred, and the temperature is adjusted at a ladle metallurgy station. It is then brought to a continuous caster, where the steel is poured into a tundish, a bathtub-shaped container. The steel in the tundish drains into an oscillating copper mold, where it produces slabs between 50 and 220 mm (2 and 9 in.) thick.

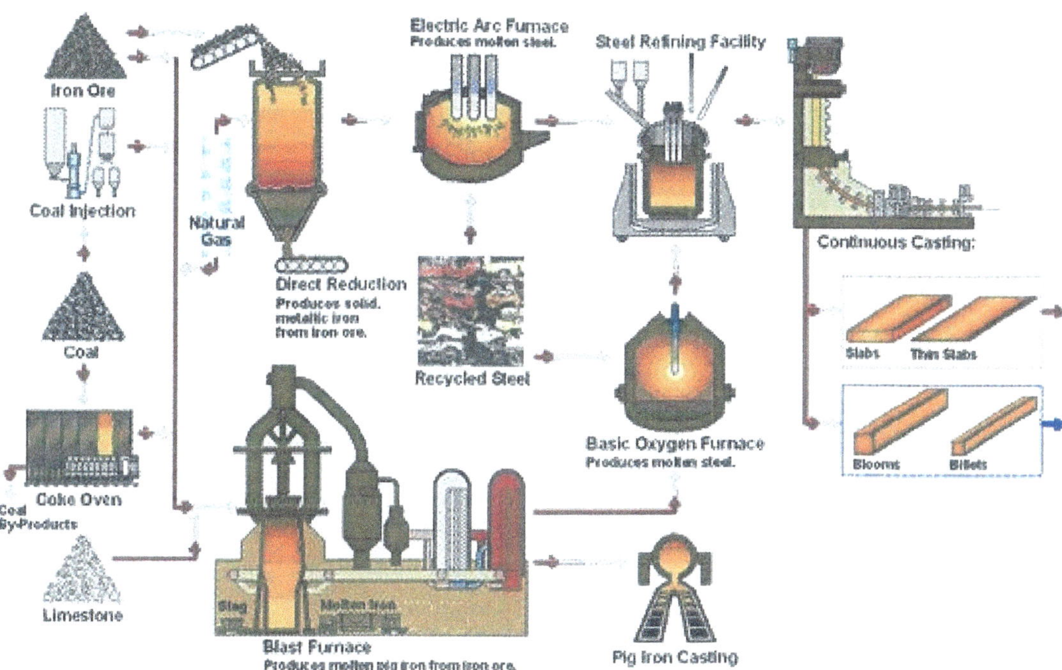

Figure 1.19 The steel-making process

The next step in steelmaking is processing the steel for the different applications (Figure 1.20 [1-7]). These slabs progress to a hot strip mill, where the slabs are rolled down to a thickness of several millimeters, but typically not below about 2 mm. At this point some of the steel is coiled for sale as hot rolled black. The steel, which is not shipped at this point, goes through pickling baths, where the scale (surface oxides formed at high temperatures) is removed. Some of this hot rolled steel is shipped to automobile manufacturers or Tiers 1 and 2 suppliers as pickled hot band to produce chassis parts. Some of the steel goes into a cold strip mill, where the steel is further reduced to steel sheet, which is typically between 0.6 mm and 2.0 mm. Since the process of cold reducing work hardens the steel, it is then heated in annealing furnaces, where it is softened. This annealing process can be either a batch annealing process or a continuous annealing process, where the desired phase relationships within the steel are established. Because there is typically better temperature control in a continuous annealing line (CAL), most AHSS must be annealed or heat treated in a CAL. The steel then moves on to a temper mill, a rolling mill, where suitable surface finish and properties are developed. At this point some of the steel (i.e., the uncoated steels) is shipped to the auto companies or their suppliers as cold rolled annealed steel to be used for uncoated (nongalvanized) body parts. Since some of the steel will be used in corrosion sensitive areas of the vehicle, that steel is sent to a coating line, where typically electro-galvanized, hot-dipped galvanized, or hot-dipped galvanneal coating is applied. There are some other coatings, but these are the primary coatings used in the auto industry. After coating, the steels coils are shipped.

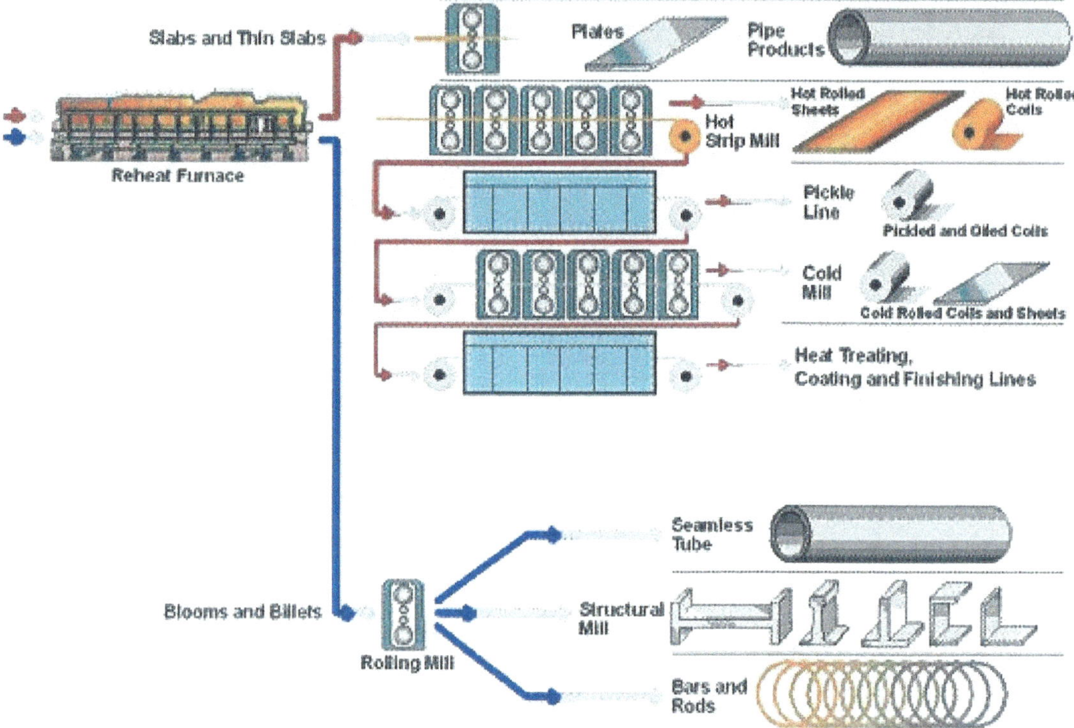

Figure 1.20 Steel processing

At this point, it will be instructive to spend some more time on the individual elements of the steelmaking process. A detailed picture of molten iron being transported from the blast furnace to the BOF is shown in Figure 1.21 [1-7]. Prior to pouring the molten iron into the BOF vessel, the molten iron is desulfurized. This is typically done by submerging magnesium-soaked coke into the molten metal. Desulfurization is done as it is poured into the BOF vessel because the sulfur levels are very high in the molten metal coming from the blast furnace, and desulfurization is more efficient and cost-effective at this stage.

Figure 1.21 Transporting molten iron from the blast furnace to the BOF

Figure 1.22 [1-7] shows scrap being added to the BOF.

Figure 1.22 Scrap being added to the BOF

Figure 1.23 [1-7] shows a depiction of a BOF vessel in operation, where metal is on the bottom and a gas-slag-metal emulsion is at the top. These two layers are separated by slag. Because of homogeneity and other concerns, it is impractical to add alloying elements directly to the BOF. Oxygen is added through a lance, which is a vertical pipe that is lowered into the BOF. Since the contents have to be moved from the BOF to the caster line, via a ladle, the whole BOF has to be tilted so the contents will come out of the taphole. Figure 1.24 [1-7] also shows the BOF in various tilted positions.

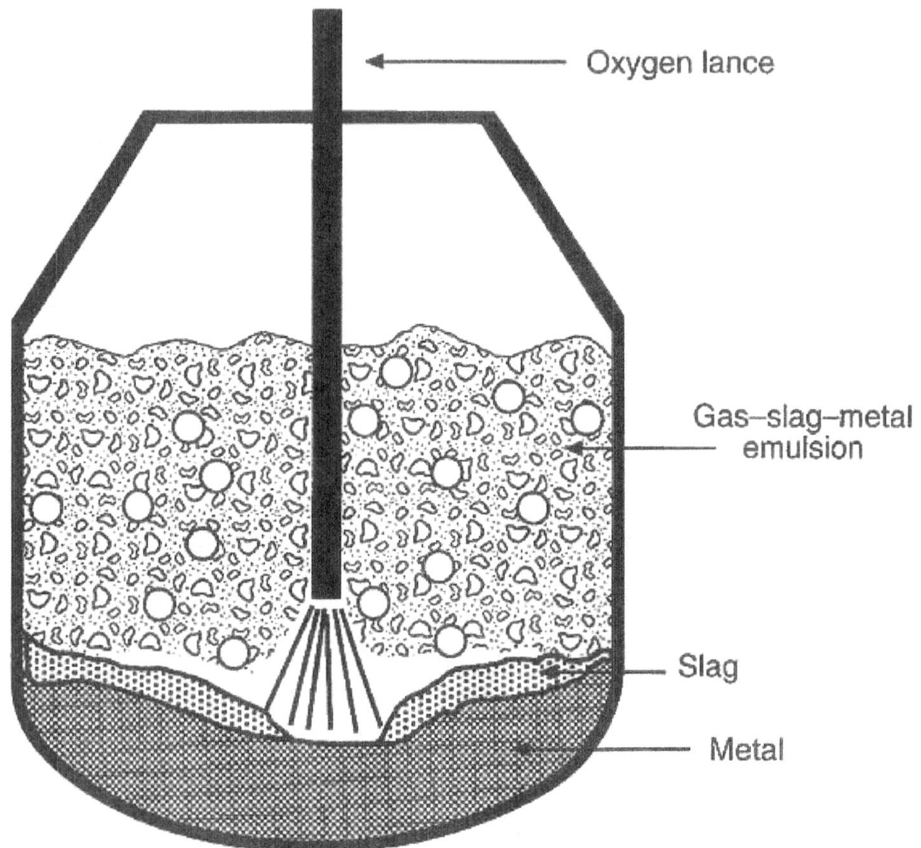

Figure 1.23 Details of the BOF

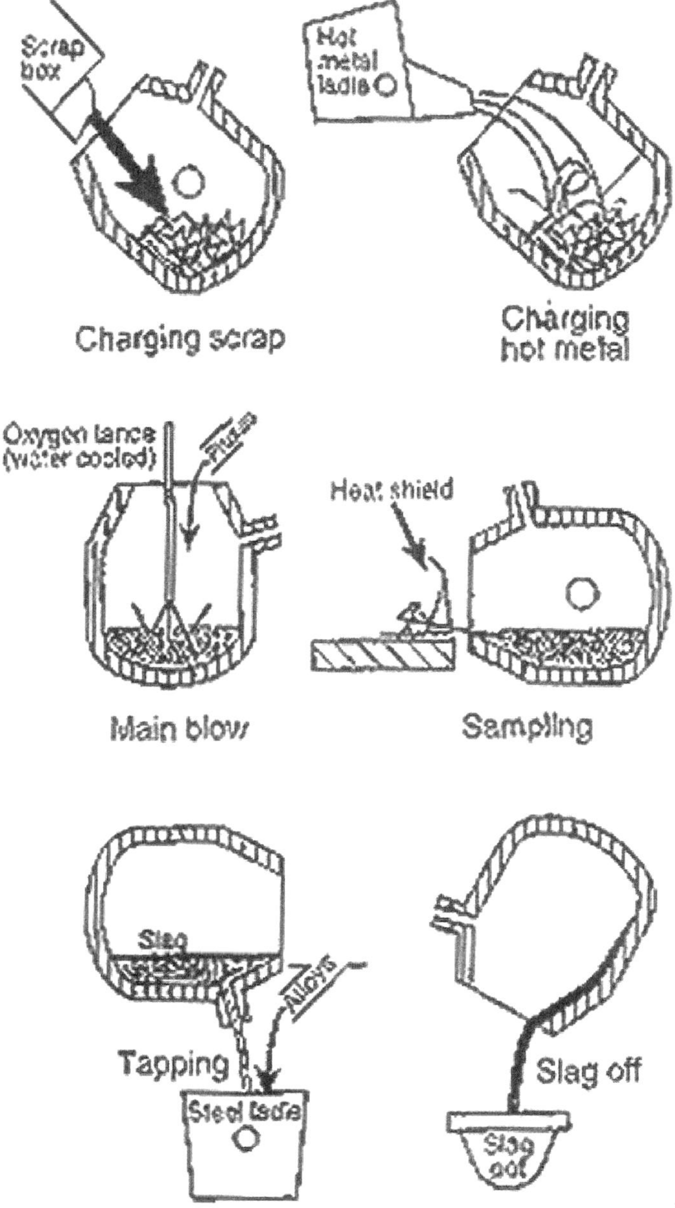

Figure 1.24 Tilting of the BOF

After the steel is tapped from the BOF into a ladle, it goes to a ladle metallurgy station, where the alloys are added (Figure 1.25 [1-7]), the steel is stirred with argon to float out inclusions and homogenize the heat, and the temperature is adjusted (typically using steel scrap as a coolant). AHSS grades contain carbon and manganese, some aluminum or silicon, microalloys such as titanium, niobium, or vanadium, and may contain other alloys such as chromium, nickel, molybdenum, or boron.

Figure 1.25 Ladle metallurgy station

The steel is now on its way to the continuous caster via a delivery ladle (Figure 1.26 [1-7]). At this point the molten steel is at about 1650 °C (3000 °F), with 98% iron, between .04 and 1.5% carbon, 1% manganese, along with other alloys.

Figure 1.26 Delivery ladle

At the continuous caster, a valve on the bottom of the ladle called a slidegate is opened and the steel is allowed to flow into a large container called a tundish (Figure 1.27). The molten metal leaves the tundish through a nozzle in the bottom and flows into a mold, where it starts to solidify. The partially solidified steel then passes between a series of rolls, with water spraying on the solidifying slab in between the rolls. The slab passes through these rolls and water spray sections while being curved to make the slab come out horizontal, after which it is cut into slabs.

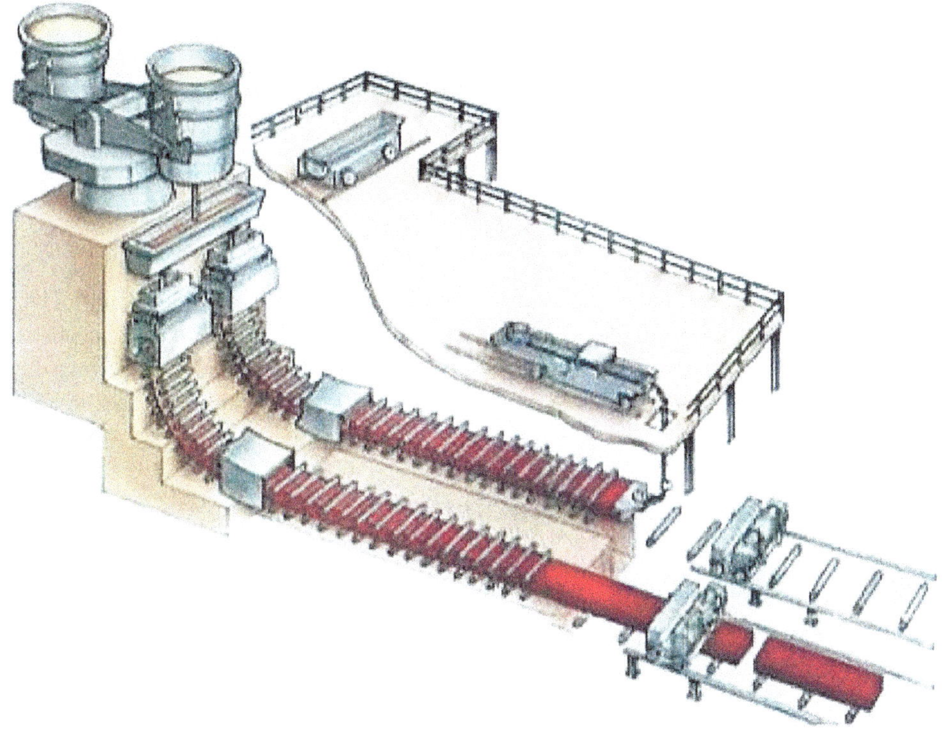

Figure 1.27 Continuous caster

Although this figure shows one ladle feeding two casters, it is typical to have two casters, each with its own tundish and ladle of steel.

An actual photograph of slabs exiting a continuous caster is shown in Figure 1.28 [1-7]. Note that the slabs are solidifying from the surface inward.

Twin Strand Caster for Sheet Production

A typical twin strand caster in an integrated steelmaking plant can produce 8,000 – 10,000 tons per day of slabs.

Figure 1.28 Photograph of continuous caster

Figure 1.29 shows the continuous caster, the hot rolling mill, and the cold rolling mill. The rolls shown on the floor next to the hot rolling mill are replacement rolls that have come back from maintenance. Rolls in the mill are frequently inspected. If they need maintenance they are sent out and replaced very quickly, because replacement rolls are always available.

Process for converting slabs to final cold rolled coils.

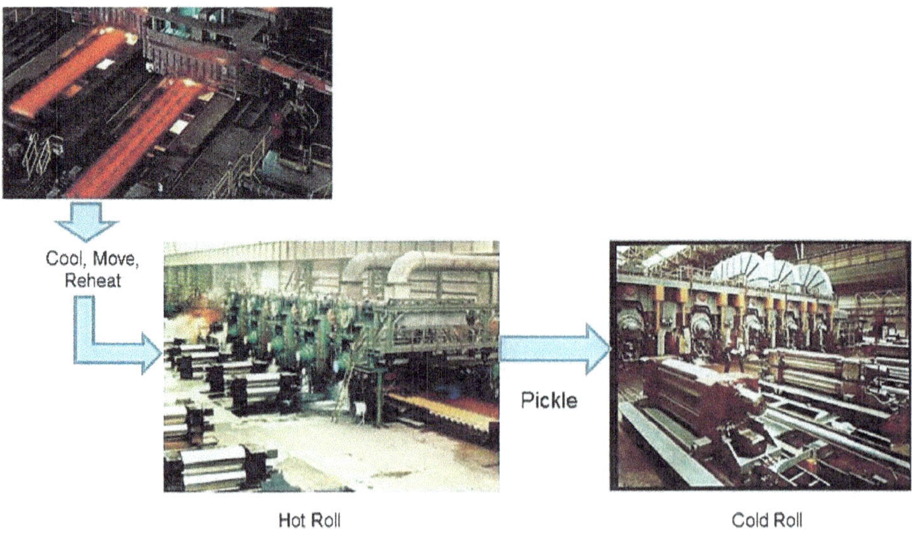

Figure 1.29 Continuous casting, hot rolling, and cold rolling

In the hot strip mill (Figure 1.30), slabs are reduced to approximately between 1.9 mm and 12.7 mm thick.

Hot Strip Mill

Slabs are rolled at starting temperatures of 2200F° and finished at 1625F°.

Slabs are reduced from 8″–10″ to thickness ranging from .075″ to .500″.

Figure 1.30 Hot strip mill

In the cold strip mill (Figure 1.31), steel sheet is reduced from thicknesses leaving the hot strip mill to between .5 and 2 mm. The steel progresses through the cold strip mill at very fast speeds. As in the hot strip mill, rolls are available to replace rolls that need maintenance. This figure shows a tandem cold mill, which has the sheet pass through multiple cold mill stands to roll the steel to the finished gauge. Some cold mills use a reversing mill, which has only one stand, where the coil of steel is fed back and forth through the mill multiple times to reach the final gauge.

Five-stand cold rolling mill reduces hot bands to customer's final thickness.

Rolling speeds modern cold mill are 4000 to 5000 fpm. Orders are sequenced according to type and size.

Figure 1.31 Cold strip mill

After passing through the cold strip mill, the coils are at their final gauge but are cold worked, making them hard and not very formable. From here, the coils can go through another thermal process to obtain their final properties. They can be batch annealed (stacks of coils heated in an inert atmosphere for about a day to recrystallize and soften the steel) or processed through a CAL, which can anneal the steel to make it soft or heat treat it to give it much higher strength. Another method of heat treating is to coat the steel in a galvanizing line, where it passes through a furnace section before being galvanized.

1.5 Coatings for Advanced High-Strength Steels

Since the coating line is the last major step in the production of automotive sheet metal and since the types of coatings specified by the automobile part manufacturer may affect the types of AHSS that can be used, this is a good time to review the basic steps of the coating process and to review the advantages and disadvantages of the various coatings. In this section, we will discuss the differences in common steel coatings from a broad product engineering perspective versus an in-depth metallurgical perspective. Also, we will address the differences from the perspective of possible changes in coatings, which might be expected in the future. Since the primary competition on coatings is between electrogalvanized (EG), hot-dipped galvanized (GI), and hot-dipped galvanneal (GA), it is important to point out the differences between these coatings. The following serves as the background for this discussion:

1. The two largest American-based auto manufacturers have traditionally used EG for exposed panels and GI for unexposed coated panels. GA has been used occasionally for thin gage unexposed panels.
2. Most of the Japanese manufacturers and one of the traditional American manufacturers are using GA extensively for both unexposed and exposed panels.
3. The Japanese auto companies are primarily using GA; and the European auto companies are primarily using a mix of EG and GI.
4. Market forces are causing Ford, GM, and others to try to minimize the amount of EG for exposed panels.
5. The limited availability of exposed quality GI is creating a dilemma as to whether GI or GA will be used as the low-cost replacement for EG.
6. All the manufacturers tend not to use coated steels if the panels are both unexposed and non-wet. That is, if water or another corrosion agent cannot get to the panel, there is no need to use coated metal for that part.

As the name implies, EG is applied by an electro-deposition process. EG is well recognized as the most robust coating for exposed automotive sheet metal applications. It adheres well to the surface of the sheet metal and provides a good surface for stamping, welding, and painting. However, EG requires a lot of electrical energy to deposit it onto the surface of the sheet metal. This explains in part why the Japanese manufacturers have chosen not to use EG and are using GA primarily, as the cost of electricity is relatively more expensive in Japan than in the United States. The desire for commonality with Japanese production and design has driven the Japanese to use GA throughout the world, regardless of where the production actually takes place. GI and GA production is done with a hot-dipping process (Figure 1.32). The only difference in GA production from GI production is that the steel goes through an annealing furnace after it comes out of the galvanizing pot. The following are relative advantages/disadvantages of GI versus GA:

- Both give acceptable corrosion protection. However, a different metric is used with each to indicate the amount of material deposited onto the steel. GI uses a "G" metric and GA uses an "A" metric: 60G and 45A are considered to be equivalent for corrosion protection and are commonly specified for GI and GA coated steels.
- Less friction equals more formability for GI. Also, GA tends to powder during stamping, which means more die maintenance. Pre-phosphating is one method of improving GA forming. The formability disadvantage is one reason why companies that use GA for exterior panels typically use more ductile grades of steel for their exteriors. For instance, they may use more IF versus BH and HSLA grades. Ironically, the Japanese are using higher grades of steel on the body interior versus their U.S. counterparts. The use of GA on exteriors may limit the AHSS applications for those panels.
- GA spot welds easier than GI, especially at the thinner gages.
  - GA requires less weld tip maintenance.
  - GA has wider weld windows, which are a measure of robustness.

    - Some would recommend not using gages less than .75 mm for GI. This would probably limit the use of AHSS for thin gage GI.
    - Improved welding equipment and different weld schedules/caps could minimize this disadvantage of GI.
- With current technology, GA and GI are nearly equivalent for painting, though painting experts might argue that point. EG is clearly superior for painting.
- Though both GI and GA are significantly less expensive than EG, GA is slightly more expensive than GI, stemming from the added process to make GA.
- There have been some issues with stone chipping for GA, which is why some have stayed away from GA for hood applications.
- GA is more brittle and the steel-zinc interface is not well defined due to alloy migration, which could cause problems when this coating is used with structural adhesives. Under some loading conditions, the adhesive/GA layer could pull away from the steel substrate, causing corrosion or loss of stiffness. GA should probably not be used in conjunction with structural adhesives in areas where high loads are encountered; or another strategy is not to use GA on panels where structural adhesive is used. This consideration is an important constraint for the use of AHSS, as structural adhesives are an important enabler for AHSS, as will be shown later in this book.

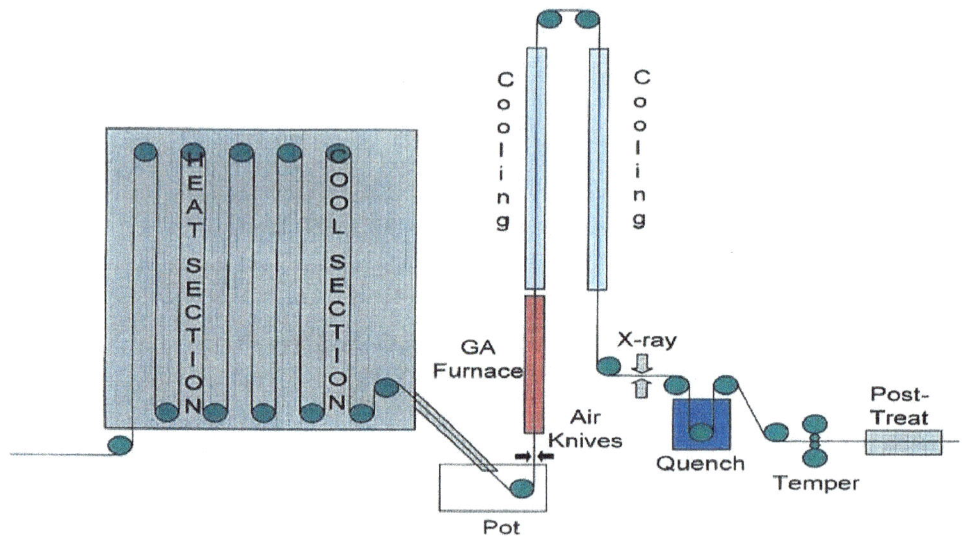

Figure 1.32 GI versus GA production

In conclusion, the following can be said:

- With appropriate manufacturing process adjustments, EG, GI, or GA can be appropriate coatings for automotive/truck usage.
- As competition among OEMs accelerates, some EG-coated panels will be replaced with GI- or GA-coated panels because of cost.
- EG will probably retained for stone chip areas (e.g., hoods) and for panels using high-strength martensitic grades, because for these grades the microstructure may be altered in the galvanizing pot.
- In the near term, forming limited panels will gravitate to GI and welding limited panels will go to GA.
- The Japanese OEM transplants will grow the production of GA in the Americas and Europe, as their processes have been optimized for GA, while the European OEM transplants in the Americas may want EG or exposed GI.
- The two largest domestic U.S. auto companies will probably use both GI and GA in conjunction with EG, but the GI to GA panel ratio will probably remain greater than one.

1.6 References

1-1. Geck, P., Goff, J., Sohmshetty, R., Laurin, K., Prater, G. Jr., Furman, V., 2007, "IMPACT Phase II—Study to Remove 25% of the Weight from a Pick-up Truck," SAE Paper No. 2007-01-1727, SAE International: Warrendale, PA.

1-2. Kim, H., McMillan, C., Keoleian, G. A., Skerlos, S. J., 2010, "Greenhouse Gas Emissions Payback for Lightweighted Vehicles Using Aluminum and High Strength-Steel," *Journal of Industrial Ecology*, Yale University.

1-3. Porsche Engineering Service, Inc., "Ultra Light Steel Auto Body Report for Phase I and Phase II Findings," March 1998, published by ULSAB Consortium.

1-4. ULSAB-AVC Consortium, 2002, *ULSAB—Advanced Vehicle Concepts*, World Steel Association, Brussels, Belgium.

1-5. Future Steel Vehicle (FSV)—Report Referenced with WorldAutoSteel Council and EDAG Inc. link available at http://www.autosteel.org/, accessed December 2012.

1-6. Denning, L., "Kafkaesque Twist in U.S.–Europe Energy Gap," *Wall Street Journal*, December 26, 2012.

1-7. American Iron and Steel Institute (AISI), link available at www.steel.org, last accessed on December 2012.

Chapter 2
Impediments and Enablers for Advanced High-Strength Steels

2.1 Forming Advanced High-Strength Steels

This chapter will cover the various impediments and enablers of advanced high-strength steels (AHSS). The presentation will proceed from the most important impediment/enabler to some of the least important. The number one impediment for using AHSS is forming problems. Even though AHSS are more formable than equivalent tensile strength conventional high-strength steels (CHSS), there are major forming problems because typically one is using higher strength AHSS than the CHSS one would be replacing. For instance, HSLA 350 would be replaced with DP 600 or DP 780, both of which would have higher tensile strengths than HSLA 350. The good news is that there has been a lot of research performed on forming AHSS, so that there are parts being used on current vehicles with as high as 1000 MPa tensile strength that are cold stamped. In addition, there are many parts with over 1000 MPa tensile strength, which are being hot formed with a subsequent increase in processing cost. Figure 2.1 shows the issues with forming AHSS (on the left side) and the countermeasures that are being applied to overcome the problems.

Forming – Stamping, Hydroforming, Roll Forming

Issues	Countermeasures *Industry Programs to address the issues*
• **Lack of forming technical data and knowledge on AHSS** • **Technical issues** - Springback - Stretch flanging - Availability of FLD - Minimum bend radius - Die wear - Improve FEA simulation accuracy	• **Formability of AHSS** — AISI-DOE Project TRP0012 • **Formability Guidelines for AHSS** Published by A/SP in 2009 • **High-Strength Steels Stamping Project** - A/SP-developing tooling practice to reduce springback - A/SP-developing FEA technique for springback • **Hydroform Materials and Lubricants** — A/SP Project • **Tribology** — A/SP project to study the surface friction, lubrication, and die wear

Figure 2.1 Forming of AHSS

Some of the technical issues are springback, stretch flanging, availability of forming limit diagrams (FLDs), minimum bend radii, die wear, shear fracture, and adequate finite-element analysis (FEA) accuracy. Springback occurs when steel is strained beyond its elastic limit and then the forming stress is released. The steel retains some of its newly formed shape. When you look at the forming strains of the part on a stress-strain diagram, you see that releasing the tensile load relaxes the steel on its new elastic strain line, which is parallel to the original elastic line (Figure 2.2). The difference between the strain at the maximum deformation point and the strain that remains after the force is released is the springback strain.

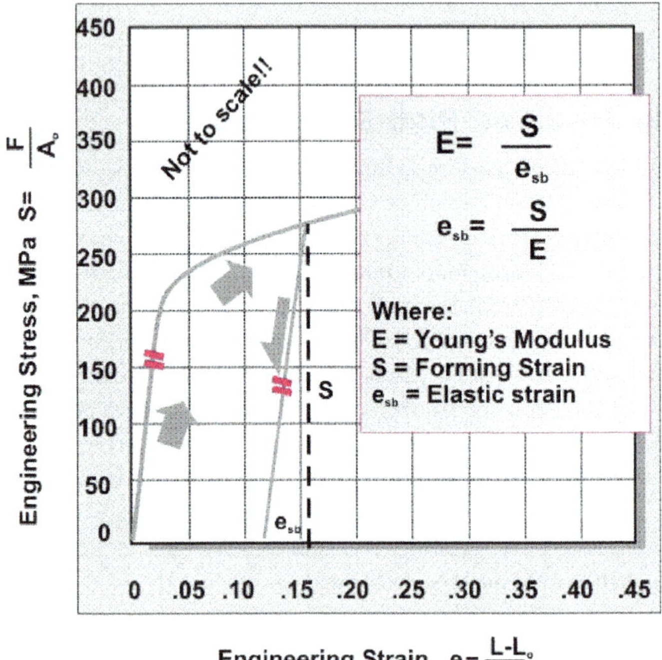

Figure 2.2 Springback

For AHSSs with a high yield stress and a high ultimate stress to yield stress ratio, this springback can be quite high. "Over-stamping" will not necessarily be possible because of die release issues. Another issue with high-strength steels in general is splitting, which can occur during stretch flanging. Sheared or stamped edges that have been stretched during forming or convex flanges (Figure 2.3) are called stretch flanges. The outside edge is where splitting may occur because this is the most highly stressed part of the flange and because sheared or punched are work hardened and have edge damage, which can be nucleation sites for crack propagation. Stretch flanges might be compensated for by using sharp punches or knives with tighter clearances, producing a very sharp, smooth edge or by using notches or scallops in the blank design. Another issue sometimes occurring with stretch flanges are bend radius issues. Too sharp a bend radius can produce cracks. Of course, a better way of fixing these problems is by altering the overall shape of the part. Strategies for doing this are demonstrated later in this book.

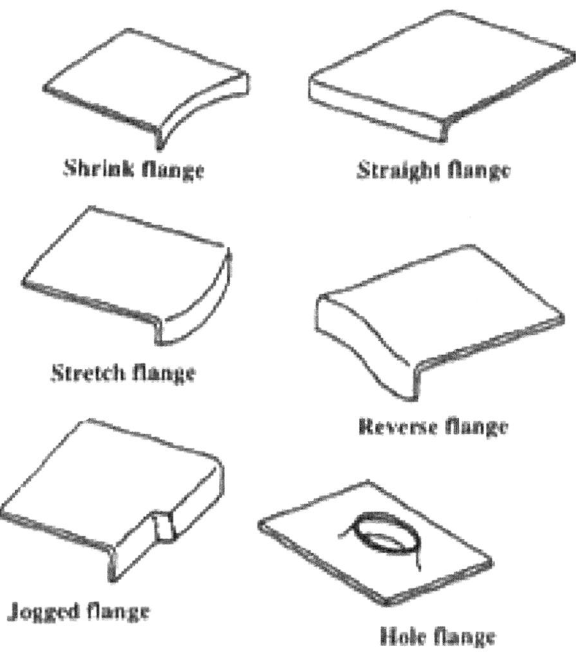

Figure 2.3 Flange configurations

The ability to form a given part is often investigated with the use of FLDs. These are generated for a specific part experimentally by etching small-diameter circles (2.5 to 5 mm) throughout the steel sheet prior to stamping, which are measured after forming. After the sheet is stamped into the desired shape, the major and minor strains (percent increase or decrease in the length and width of the circle) are plotted for each of the circles on a FLD (Figure 2.4). Two curves are generally plotted on the FLD. These are the failure limit curves (FLCs). The upper curve defines when necking occurs at a point on the part. If a point on the sheet metal is plotted above the top curve, a failure in terms of a tear can be expected. Normally, a lower curve is plotted, which may be 10% below the top curve. The distance between the two curves defines a yellow zone in which failure might occur. If the point is below the bottom curve, the point is considered safe and no problem can be expected. There is a formula that defines the FLC for most mild, medium, and high-strength steels, which is a function of the n-value and the thickness for the part [2-1]. However, for AHSS, this formula is not very accurate in predicting failure, especially for positive minor strains. This is because a new failure mechanism known as shear fracture comes into play. Shear fracture is a complex failure mechanism that is currently addressed by limiting the use of sharp bend radii in drawn sections. Since the FLD cannot predict this failure mechanism, design engineers need to understand the way the steel flows in the die to prevent heavy stretching over a tight radius.

It is helpful if the steel manufacturer supplies the FLC for the steel that they are supplying by generating the forming limit curve from physical testing. However, the FLC may not be available, as experimental determination of the FLC is quite difficult. Also, the FLC for two different DP 600s, produced by two different manufacturers, might vary as different procedures might be employed (e.g., chemistry intensive versus process intensive) in making the steel. This presents an impediment for utilizing AHSS. One benefit to FLD production is the fact that computer-aided engineering (CAE) models are fairly effective in producing FLDs, eliminating the need to build a soft tool (e.g., Kirksite) and stamping a bunch of parts with circle grids etched on them. This brings up another impediment of using AHSS in that soft tools could experience failures (cracks), as the strength of the steel being stamped can be very high.

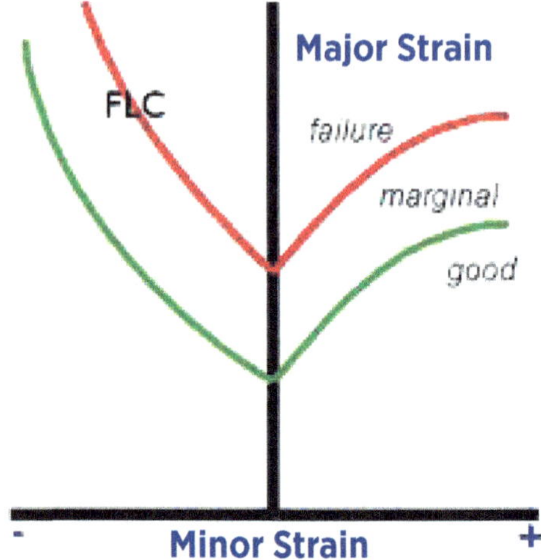

Figure 2.4 Forming limit diagram

Usage of AHSS has been a springboard for the use of CAE in stamping simulation for the previously stated reasons. However, there is a significant learning curve in using simplified one-step stamping codes (e.g., FAST_STAMP) and for incremental codes (e.g., LS-DYNA). The basic difference between these two types of codes is that with the incremental code the stamping event is simulated from the flat steel sheet to the final part, whereas with the one-step code the FEA starts with the finished part and pushes the metal into a flat sheet (which is the reverse of the incremental approach). The effort in model building and solution times can be significantly reduced by using one-step codes, but with much less accuracy. One-step codes are typically used to give an approximate answer, and for parts that show problems, this analysis is followed up with an incremental analysis.

To counteract these impediments, the steel and automotive industries (e.g., American Iron and Steel Institute [AISI], Auto/Steel Partnership [A/SP]) along with some support from the U.S. government (the Department of Energy [DOE]) have been performing research in all of the discussed areas. On the right side of Figure 2.1, several areas of research are identified. AISI-DOE Project TRP0012 [2-1] is a very complete report on forming AHSS. Formability guidelines are covered in the second study [2-2], which contains a number of case studies compiled by the A/SP. The other studies listed are ongoing projects at the A/SP. A good overview of hydroforming AHSS was presented in 2012 at the Great Designs in Steel conference [2-3].

In the next couple of figures, some of the forming guidelines, which were an outgrowth of the Improved Materials & Powertrain Concepts for 21st Century Trucks (IMPACT) Project, are listed (Figures 2.5 and 2.6).

Risk Assessment — Forming

- **Symmetrical parts with constant cross-section w/o bends**
  - ⇒Roll formed
    - Process variables well established
    - Guidelines similar to UHSS
      - Used over several decades for door beams, bumpers, etc.
- **Symmetrical parts with bends**
  - ⇒Conventional forming
    - 4-piece die (*vis a vie* 3-piece)
      - Until experience gained
    - Springback issues similar to conventional HSS

Figure 2.5 Forming symmetrical AHSS parts

Risk Assessment — Forming

- **Asymmetrical parts with AHSS**
  - ⇒ Conventional forming
    - Springback, twist extremely difficult to control
  - ⇒ Engineer the die process to accommodate AHSS
    - 4-piece die = norm
    - Distribute the strains to control twist
    - Engineer stiffening features to control twist and springback

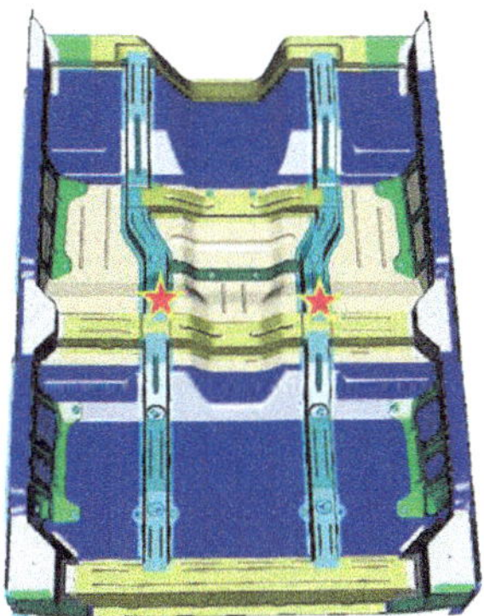

Greater risk, but manageable

Figure 2.6 Forming asymmetrical AHSS parts

Symmetrical parts with constant cross section, without bends, can be roll formed using very high strength AHSS. Roll forming is a process of forming a part by passing the sheet metal blank through a series of rollers, which gradually form the part by bending the sheet metal a little bit more with every new set of rollers. For symmetrical parts with bends, four-piece dies may be required. Figure 2.7 shows a two-piece die for uncomplicated, mild steel parts [2-4]. Figures 2.8 and 2.9 show two different examples of three-piece dies, which give more control to the build-up of stresses in different areas of the part, and Figure 2.10 shows a four-piece die, which gives even more control over stress build-up. With asymmetrical parts made from AHSS, four-piece dies should be the norm. Engineering stiffening features (e.g., darts) are also used with asymmetrical parts to control twist and springback.

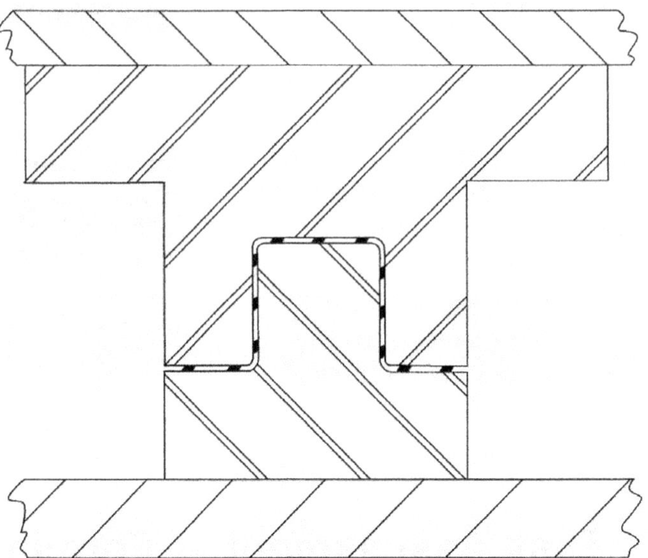

Figure 2.7 Two-piece die

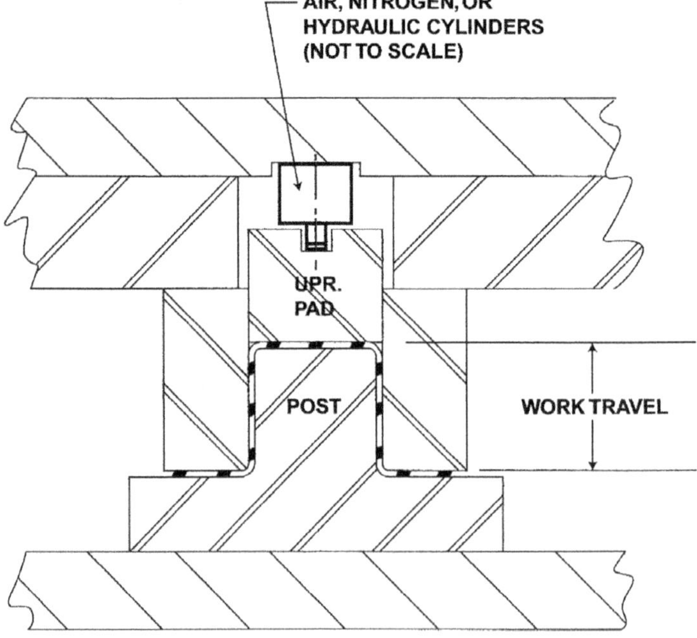

Figure 2.8 Three-piece die with an upper pad

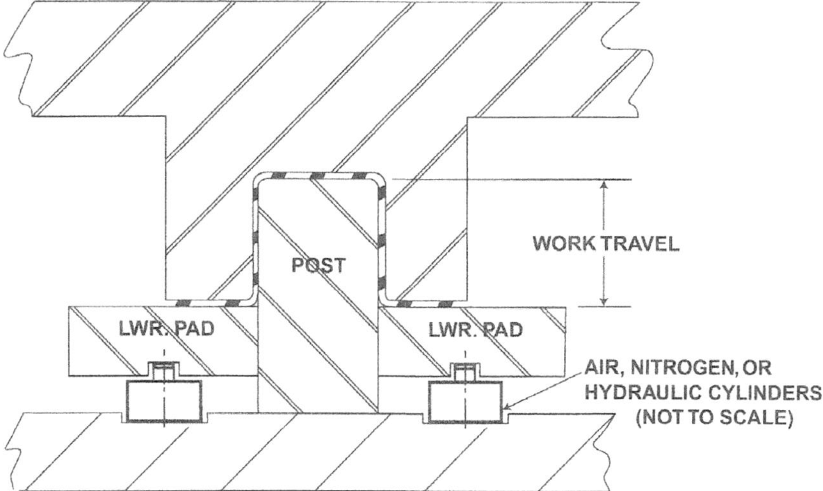

Figure 2.9 Three-piece die with a lower pad

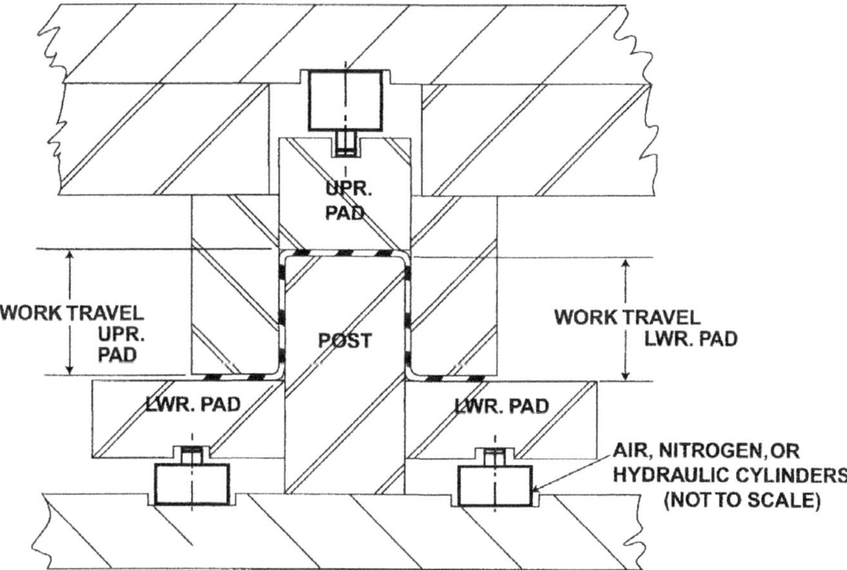

Figure 2.10 Four-piece die

Figure 2.11 shows anticipated time delays in forming development, which can occur with AHSS usage. Another contributor to time delays and to increased cost with AHSS is that die materials, for both die substrates and coatings, may need to be enhanced due to potential for die breakage and wear. There are a number of good presentations on this subject that have appeared in the GDIS conference [2-5] which you can refer to, as this is an evolving area of concern. The author believes that ultimately for very high strength AHSS, the dies themselves will have to be analyzed using FEA to take some of the art out of die architecture and material selection.

Forming Development Time Effects

Process / Part Complexity	Additional Measures – beyond current practice of one step for all new parts	Additional Development Time Required
Simple – Roll Form	None – ensure min. bend radius capability matches process requirement	None – use existing tools
Bend Only	Match min.bend radius capability to process Springback modeling (proper over bend) Darts to minimize springback. Counter measures are similar to HSS.	Minimal Delay – likely None
Closure Panels	Attention to trim development, stretch flange situation. Recent trials and studies have identified the issues and solutions proven.	2 Weeks
Symmetrical Structural Parts	Springback modeling, control darts, min. bend radius, controllable binder to stretch. Avoid restrike. 4-piece die preferred over 3-piece.	1 Month
Asymmetrical Structural Parts	Springback modeling, control with darts, Understanding the cause of twist. Avoid restrike. 4-piece die should be used directly to minimize risk.	2-3 Months

Figure 2.11 Forming development time effects

In summary, Figure 2.12 shows the major strategies that stampers have to put in place to accommodate AHSS. Another new development in stamping is the increased use of servo-presses. These presses allow the material to form better so they can get better shapes out of high-strength parts, possibly using a lower-forming grade for some applications, while achieving better dimensional accuracy. In addition, these presses are much easier on the dies, reducing fatigue, wear, and energy consumption.

AHSS Forming Implementation Strategy

- **Formulate complete forming rules**
- **Develop complete springback rules**
- **Validate FEA formability models**
- **Optimize coatings for tool wear**
- **Educate in-house, Tier 1 stampers**

Figure 2.12 AHSS forming implementation strategy

Next are some examples from some actual stamping development work that took place on the IMPACT Project. In this project, we were trying to convert some body panels that were being made from mild steel or lower strength CHSS to AHSS (Figure 2.13).

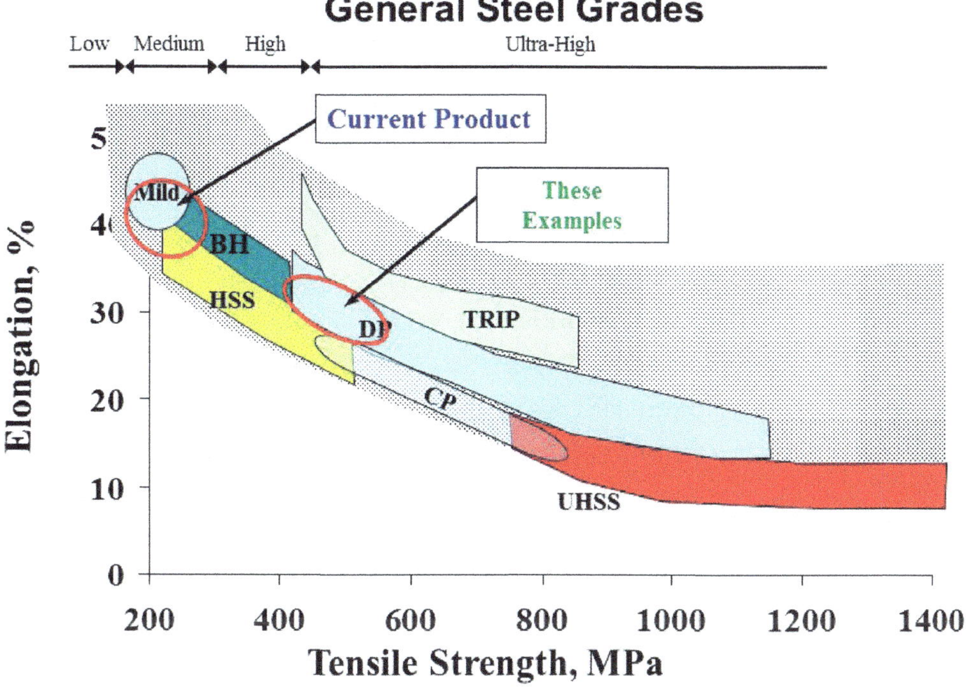

Figure 2.13 IMPACT material conversion strategy

The parts that were selected for detailed analysis in the IMPACT Project were the fender, door outer, mid-transverse floor cross-member, and the front-end longitudinal member, typically called the shot gun.

Rationale for Selection of Parts

- Fender, Door Outer
  - Customer appeal and functional performance
  - Degree of difficulty in tool buy-out process
  - Severity of post-forming operations

- Mid-transverse cross member, Reinforcement front fender inner (Shot gun)
  - Target for improving crash performance
  - Target for down gauging

Figure 2.14 Parts selected for study

I will review the fender and the mid-transverse cross-member study here. The rationale for the parts selection is shown in Figure 2.15.

Steels used in the study

- Fender
  - 0.80 mm BH210
  - 0.65 mm DP500
- Mid-transverse X member
  - 1.50 mm HSLA350
  - 1.50 mm DP600
  - 1.40 mm DP600

Figure 2.15 Rationale for parts selection

For the fender, we were trying to convert .8 mm BH210 to .65 mm DP500 (Figure 2.16). The incremental FEA results indicated several problem areas on the front fender (Figure 2.17), using DP500.

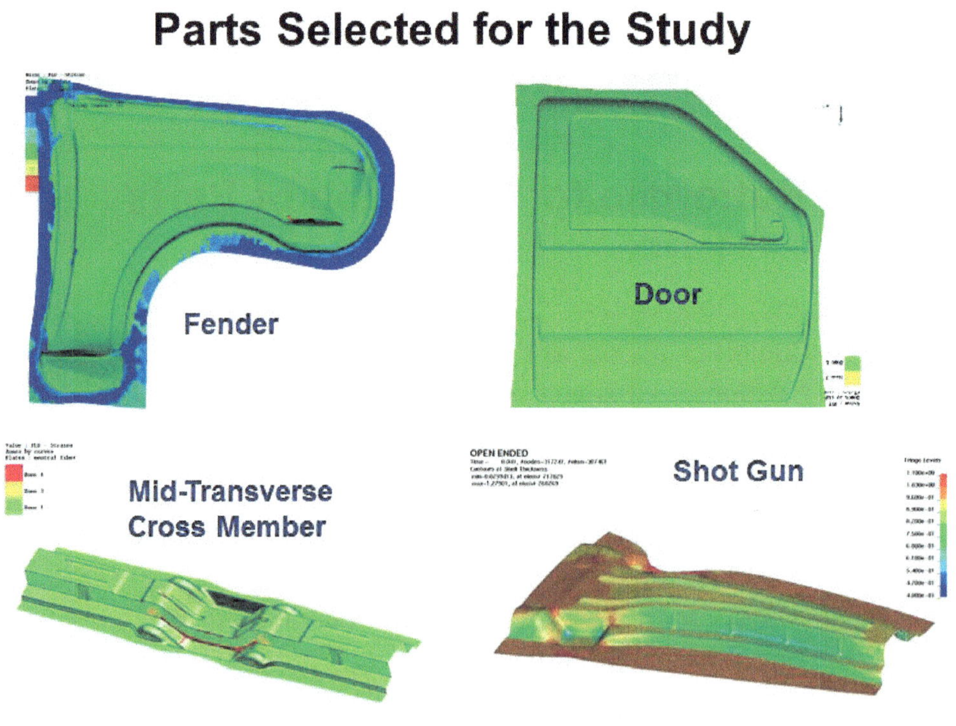

Parts Selected for the Study

Fender

Door

Mid-Transverse Cross Member

OPEN ENDED

Shot Gun

Figure 2.16 Steels used for fender study

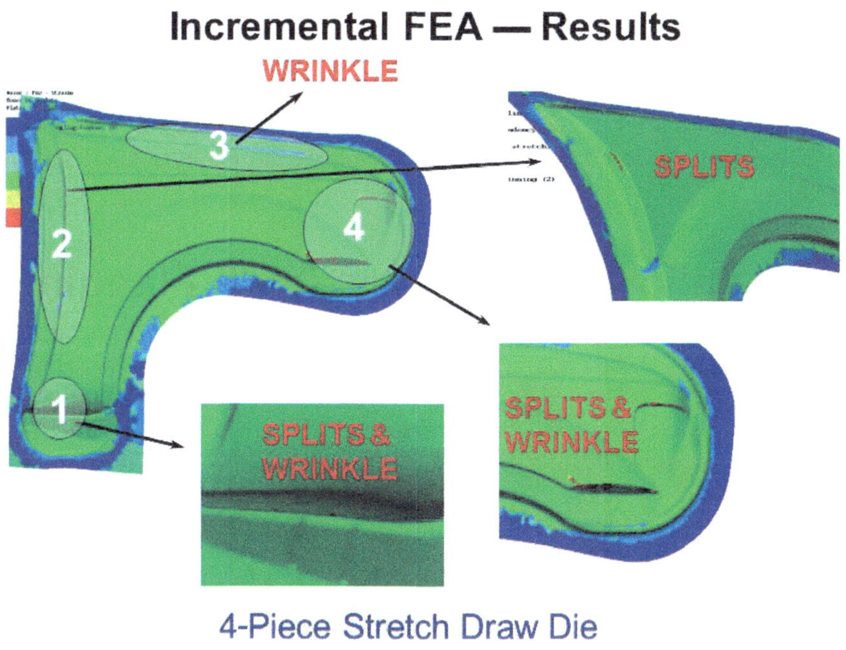

Figure 2.17 Incremental FEA results for the fender

A soft tool (Kirksite), made with a zinc-based alloy, was constructed, and the soft tool stampings replicated the FEA results (Figure 2.18).

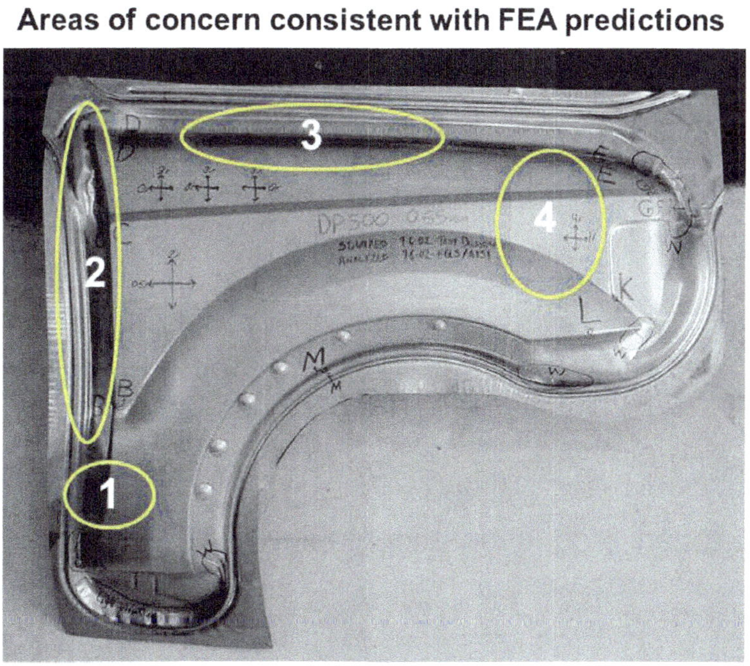

Figure 2.18 Soft tool DP500 fender stampings

Based on the FEA and soft tool results, the fender was redesigned to improve the formability (Figure 2.19) without changing the cosmetic appearance of the fender. The soft tool was then modified consistent with these changes, and an FLD was produced from the stamped part. The fender in DP500 was then judged to be acceptable. The FLD shows only a few points slightly above the marginal line, which was consistent with the quality of the produced part (Figure 2.20).

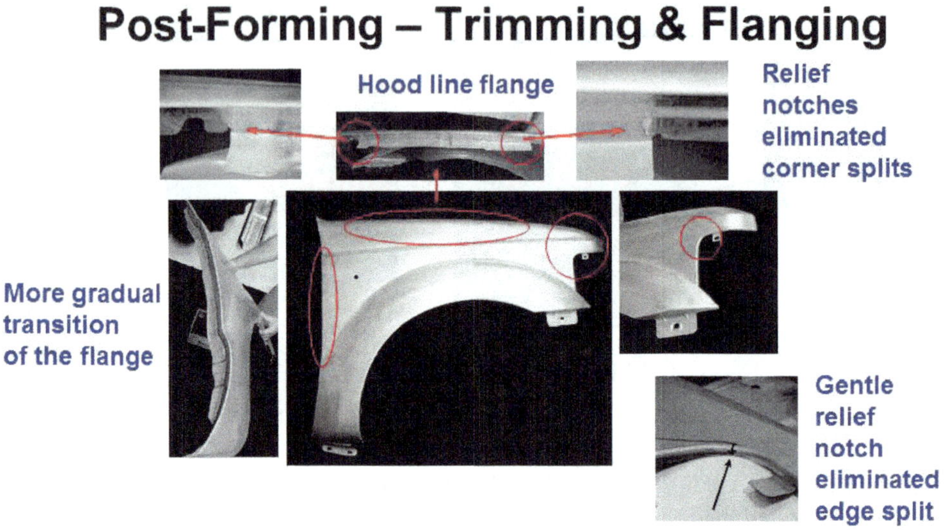

Figure 2.19 Modifications from the original design

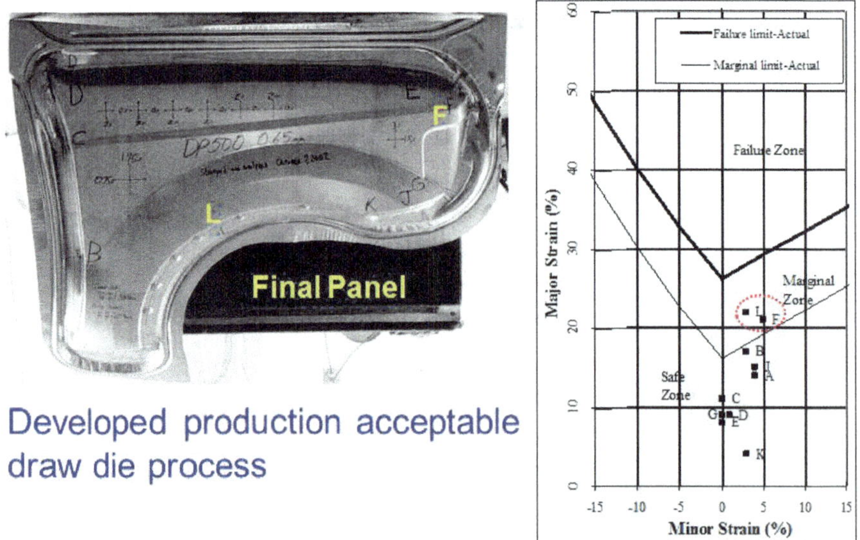

Figure 2.20 Acceptable fender after design modifications

Figure 2.21 itemizes the lessons learned from the IMPACT fender study. It should be pointed out that flanging was not done with production type flanging tools for this study. However, in some cases, the number one impediment to the use of DP 500 for exterior panels has been the potential for cracks during flanging. The number two impediment has been that many of the companies that have considered using DP 500 for body exterior panels are using galvanneal coatings, which are harder to stamp than electrogalvanized [EG]. Another impediment to using DP 500 on exteriors has been availability of panels wide enough in DP 500. This last impediment has limited DP 500 for hood applications, an ideal application, where running the blanks lateral to the hood (which is not usually width limited) creates too much offal. Because of these impediments, the result has been that only a couple of exterior panels have made it to production in DP 500 (front door panels on two Ford vehicles). We will discuss DP 500 for exterior panels later in this book.

Lessons Learned — Fender

- 0.65 mm DP500 product showed good potential for a production application with the following considerations;
  - Clean edge condition after trimming is important
  - Avoid notches in the door line flange
  - Gradual transition of flange length
  - Stretch flange conditions
    - Minimize the trim flange length
    - Eliminate sharp notches in the flange

Figure 2.21 Lesson learned from the IMPACT fender

The steels, which were studied for the mid-transverse cross-member, were 1.50 mm HSLA 350 (baseline), 1.50 mm DP 600, and 1.40 mm DP 600 (Figure 2.22). This is a part that manufacturers are trying to increase the strength on, because it is one of the controlling members for side impact crash performance. Since DP 600 has higher ultimate tensile strength than HSLA 350, there would be some utility in being able to make the part at the same gauge in stronger steel. However, since the primary goal of the IMPACT Project was to reduce weight, it was hoped that the reduced thickness panel would be achievable.

Steels used in the study

- Fender
  - 0.80 mm BH210
  - 0.65 mm DP500
- Mid-transverse X member
  - 1.50 mm HSLA350
  - 1.50 mm DP600
  - 1.40 mm DP600

Figure 2.22 Steels used for the mid-transverse cross-member

Figure 2.23 shows the incremental FEA results for the mid-transverse cross-member. At the top left is shown the strain map superimposed on the part for the 1.40 mm DP 600. The lower center shows the thickness contour map from the FEA results. At the right we see the FLD results. Most of the points on the piece are green, which is the no-problem area of the FLD. However, the edges are red for the strain on the part and the FLD. Also, much of the thickness contour map is red. Overall, one would expect wrinkling and maybe some splits. However, the soft tool–produced part (Figure 2.24) shows a fair amount of wrinkling but no splitting. It is interesting that FEA or actual part-generated FLDs may not be definitive in predicting failure modes on the part.

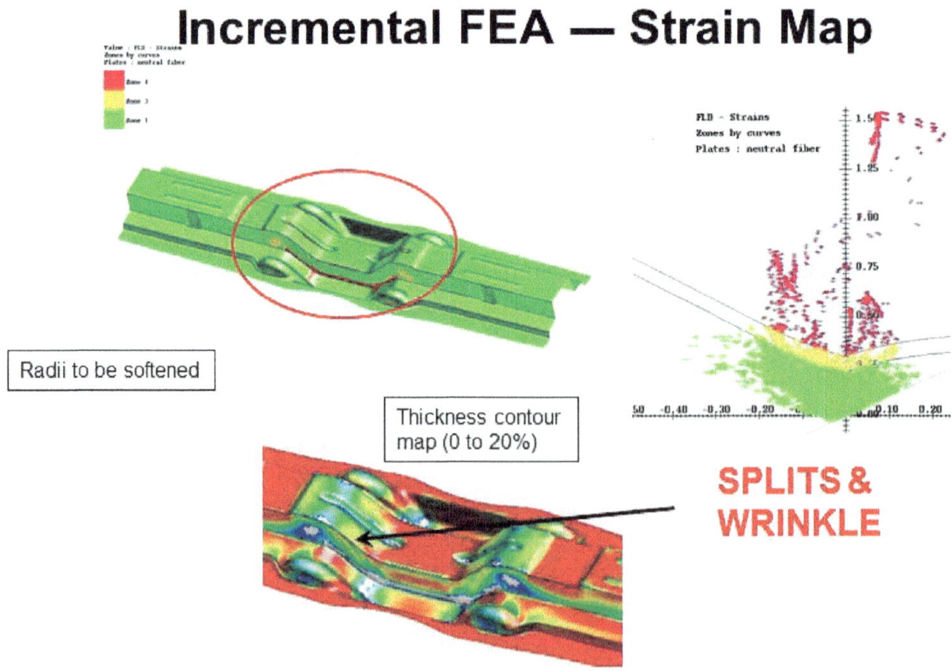

Figure 2.23 Incremental FEA for the mid-transverse cross-member

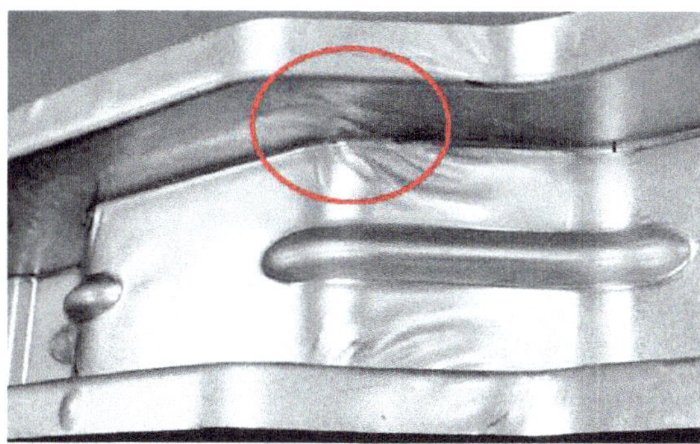

Figure 2.24 1.40 mm DP 600 mid-transverse cross-member

Figure 2.25 demonstrates the reason for the wrinkling. There is a radical change in the length of lines transverse to the center line of the part as one progresses longitudinally along the part. This is one of the main causes of part wrinkling. The part was then redesigned to reduce this problem (Figure 2.26), and the problem was reduced to an acceptable level. Measurements were also made of the amount of springback along the length of the part (Figure 2.27).

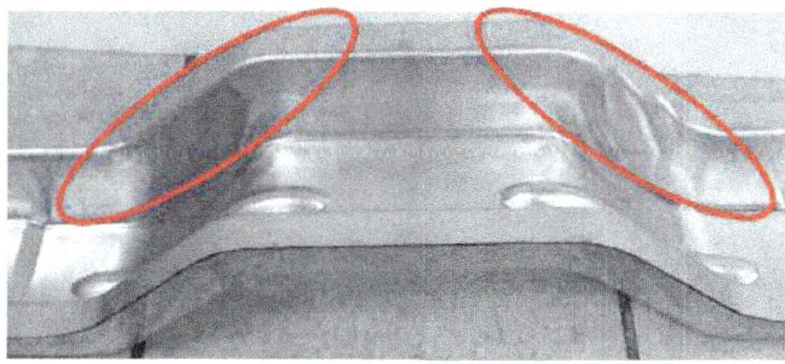

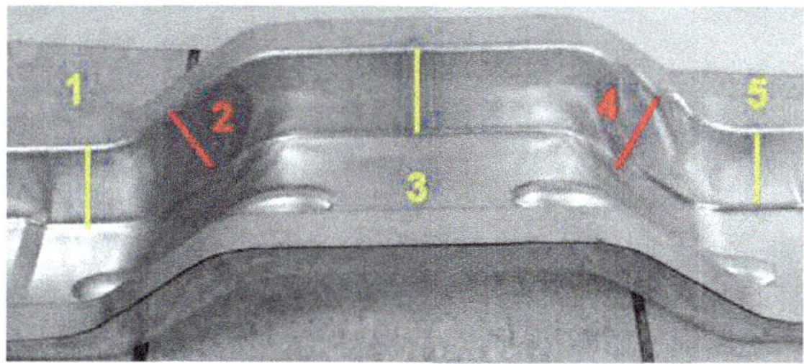

Figure 2.25 Length of line problem

Modified Tunnel Design

Constant length of line across the bend

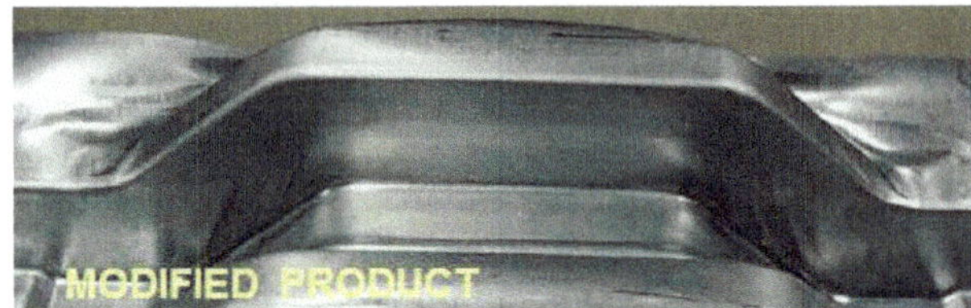

Figure 2.26 Redesigned mid-transverse cross-member

Springback — Data

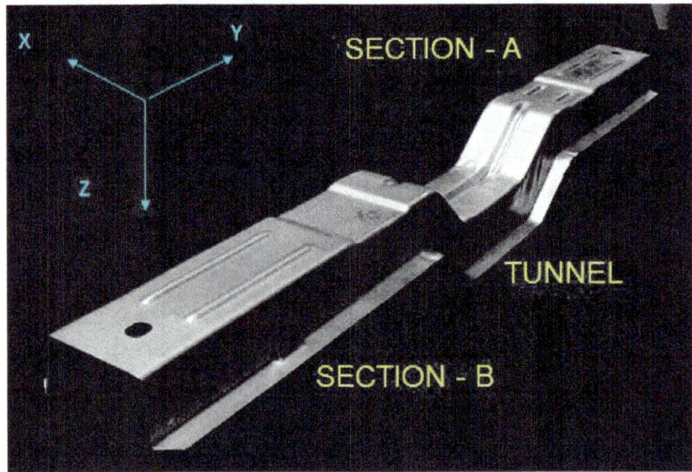

Figure 2.27 Springback measurement locations

There is slightly more springback though for both the 1.50 and 1.40 mm DP 600 versus the baseline (Figure 2.28). The springback was in the acceptable range.

Springback - Tunnel

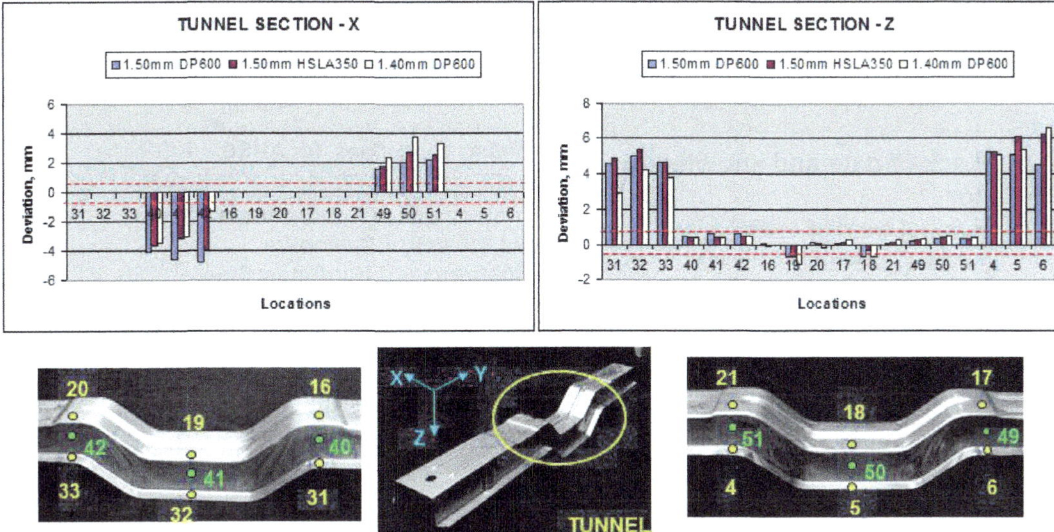

Figure 2.28 Springback results

2.2 Welding Advanced High-Strength Steels

The second largest impediment to the use of AHSS is welding, though it is not nearly as large of an impediment as formability. The issues and countermeasures with welding AHSS are shown in Figure 2.29. Part of the problem is that there is a lack of data and knowledge related to welding of AHSS, and this is true for spot welding, laser welding, gas metal arc welding (GMAW), and hybrid welding technology. Most of the welding on the automotive body is spot welding, except for the use of laser-welded blanks, which are typically welded at suppliers to the original equipment manufacturers (OEMs). This has been the case for many years because spot welding is less expensive and faster to apply than other joining methods, and it is easier to detect whether the welds are good in the body shop versus for instance GMAW or laser welding. However, for thicker gauge steel (e.g., automotive frames and subframes), GMAW welding is usually the choice. Since most sheet metal welding is resistance spot welding, we will concentrate on that area in this discussion. The primary technical issues for spot welding are interfacial fracture, narrower current range, electrode life, use of AC power supplies, and reasonable acceptance criteria.

Welding

Issues	Countermeasures *Industry Programs to address the issues*
• **Lack of data and knowledge for** - spot weld of AHSS - laser welding - hybrid welding technology • **Technical issues for spot welding** - Interfacial fracture - Narrower current range - Electrode life - Use of DC power source - Acceptance criteria	• **Welding Guidelines for AHSS** – A/SP project • **Development of Appropriate resistance Spot Welding Practice for Transformation Hardened Steel** – AISI-EWI 9934 project - COMPLETE IN 2004 • **Laser Assisted Arc Welding for AHSS** AISI-INEEL Project • **Develop Weld Parameters and Acceptance Criteria for AHSS** – A/SP Project

Figure 2.29 Welding impediments

When mild steels and CHSSs are spot welded, the resulting weldment, which is a little cylinder that is about 6 mm in diameter and is the thickness of the two (or more) sheets of steel being joined, is typically about the same strength as the parent material. When weld test specimens are made to investigate the strength of the joint (Figure 2.30), they are typically pulled apart in a tensile-test machine. However, an attempt to replicate this procedure in the body shop is done with pry-bar testing. Normally the weldment, shown as gray in Figure 2.30, will stay intact and will pull out of either the upper or the lower piece of sheet metal in the figure. This is because the weakest metal is the annular ring, which surrounds the weldment and is called the heat-affected zone (HAZ). The term used for this is "to pull a nugget." When a nugget is pulled intact, the assumption is made that the weld is good. If the nugget experiences a shear failure and part or the entire weld nugget fractures at the weld centerline, the assumption is made that the weld is bad, because the nugget was weaker than the HAZ. With AHSS, since the nugget is annealed during the welding process, the nugget may be weaker than the parent material and weaker than the HAZ. However, the weld might still be acceptable, even though there is a shear failure of the nugget itself. This creates the issue of how to test and approve welds that have been made with AHSS. One solution that is being used by several OEMs is to use a coupon setup that puts more tensile loading on the nugget versus shear loading. There are also procedures that can be used in the body shop to replicate this procedure. Ultrasonic weld checking has been used by some. Of course, having a more robust welding procedure might negate some of the coupon and in-plant testing that needs to be done. This problem has not been completely resolved, and auto manufacturers are still looking for new acceptance criteria to use with AHSS welding, especially at higher strengths.

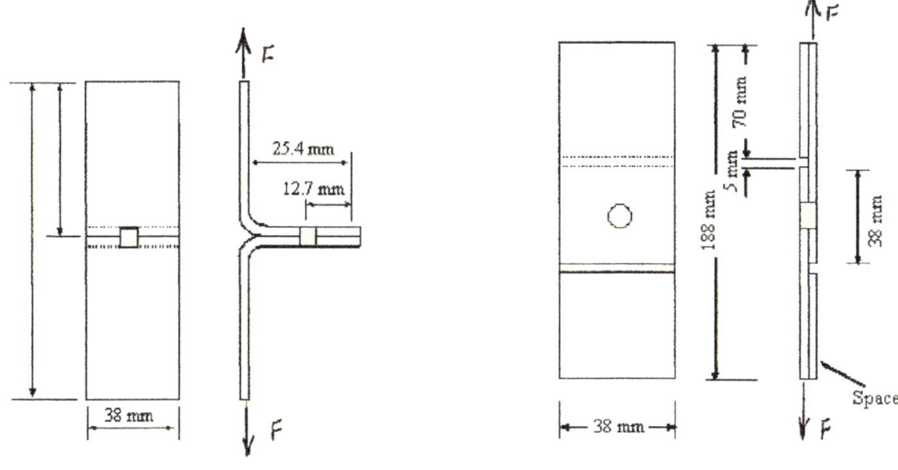

Schematic of coach peel specimen

Schematic of tensile shear specimen

Figure 2.30 Weld specimens

Another issue with AHSS is the acceptable current range or "weld window," which defines the weld parameters needed to make a good weld. When a company determines that there may be difficulties with welding a certain thickness, grade, coating, or stackup of two or more types of steel, weld coupons are generated and welded at a number of different current levels. The lowest current where a good weld is produced is recorded. The highest current level where expulsion just starts to appear is also recorded. Expulsion occurs when sparks are given off the weld, and typically beyond that point a good weld will not occur. The difference between the lowest expulsion current and the initial good weld current defines the weld window. Usually, manufacturers want this window to be greater than 1.5 kilo amps for a robust welding process. There are a number of different strategies for increasing the weld window, for instance, running more or less current cycles or using different weld tip geometries.

Electrode life has been a problem with higher strength steels in general, which could be anticipated because of the wear and tear on the weld tips (caps). Manufacturers usually want to have greater than 1500 cycles before redressing weld caps. There have been a number of investigations with different weld tip shapes. Some have found that hemispherical shapes work well, but there are a number of different shapes that work well under different conditions.

Probably the biggest innovation for welding AHSS has been the switch over from AC transformers to mid-frequency DC transformers. Fifteen years ago, almost all the U.S. automotive plants were using AC, but now concurrent with major model changeover, almost all the new transformers are DC. The DC transformers are more expensive, but besides being better for AHSS welding, there are some plant flexibility benefits. An example, which shows the benefits of mid-frequency DC versus AC, is shown in Figure 2.31. In this figure, widely dissimilar thicknesses (which by itself can be a problem) are welded with both AC and DC. The problem with the AC welding, which is exacerbated because one of the sheets of steel is AHSS, is that the nugget does not achieve good penetration into the thinner steel sheet. This problem is overcome with the DC welding, where the nugget achieves good penetration into both steel sheets.

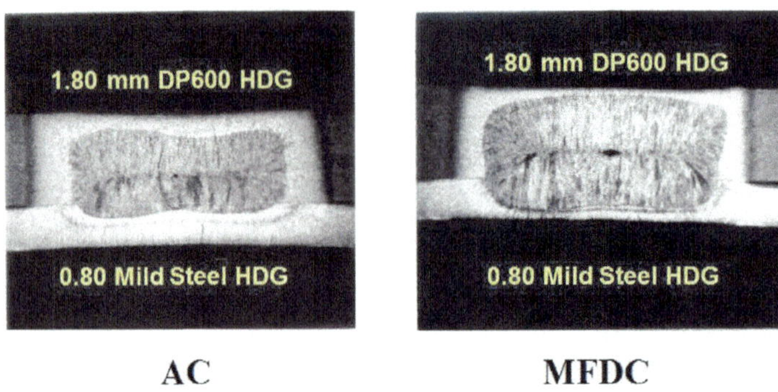

Figure 2.31 AC versus MFDC welding

Some of the studies that have been undertaken to improve AHSS welding are listed in Figure 2.29. The most comprehensive study that has been done on AHSS spot welding [2-6] can be found on www.autosteel.org.

Again, the IMPACT study provides a good example on how one might approach suspected weld problems in a new vehicle program. Figure 2.32 lists all the weld situations that might have been problematic. Several of the cases in this table reveal either current range or life problems.

Screening Study – AC Welding

No.	Material A	Material B	Current Range (kA)	Life Test
1	0.85 mm DP EG	1.5 mm BH 250 EG	3.6	1750
2	0.85 mm DP EG	1.1 mm BH 280 EG	3.8	2000
3	1.1 MM BH 280 EG	0.7 MM DP CR	2.9	2000
4	0.8 mm BH 210 EG	0.65 MM BH 210 EG	3.2	2000
5	0.65 MM IF HD	0.65 MM DP EG	2.4	750
6	0.9 MM BH 210 EG	1.3 MM DP HD	1.4	1750
7	1.2 MM DP HDG	1.3 MM DP HD	1.1	750
8	1.3 MM DP HD	0.65 MM BH 210 EG	2.0	1000
9	1.0 MM BH 280 EG	0.65 MM BH 210 EG	3.2	2000
10	1.5 MM BH 250 HD	0.65 MM BH 210 EG	1.4	1250
11	1.5 MM BH 250 HD	0.85 MM BH 210 EG	1.7	2000
12	1.5 MM BH 250 HD	0.7 MM DP CR	2.0	2000
13	1.5 MM BH 250 HD	1.0 MM BH 280 EG	3.3	2000
14	0.65 MM DP EG	0.65 MM DP EG	2.0	2000
15	0.9 MM DQ EG	0.7 MM DP EG	1.3	1750
16	0.85 mm DP HDG	0.95 mm DP HDG	2.3	750
17	0.65 mm DP EG	0.65 MM DP EG	1.3	2000
18	0.65 mm DP EG	0.85 mm BH 210 EG	1.1	1500
19	0.7 mm DP EG	0.7 mm DP EG	2.4	2000
20	0.65 mm DP EG	1.3 MM DP EG	2.2	1250
21	0.85 mm DP EG	0.65 mm BH 210 EG	3.8	1500
22	1.3 MM DP HD	1.3 MM DP HD	1.3	1750

Potential issues with these grade and gage combinations

CR = cold rolled or uncoated steel
EG = electo-galvanized coated steel
HD or GI = hot-dipped galvanized coating
HDG or GA = hot-dipped galvanealed coating

Figure 2.32 IMPACT weld screening study

Some of these combinations are displayed in a comparison between AC and DC welding (Figure 2.33). In all but one of these examples the DC welding performed better than the AC welding. One of the stackups (which was not shown in the table) is a three-thickness weld, which can be particularly problematic (Figure 2.34). For this example, increasing the weld time over what would be used for HSLA solves the insufficient weld window problem. Two of the examples from the table in Figure 2.32 are shown in Figure 2.35. It typically is more difficult to achieve 1.5 kilo amp weld windows with widely dissimilar sheet thicknesses. However, by optimizing the weld time, adequate solutions were found in both cases.

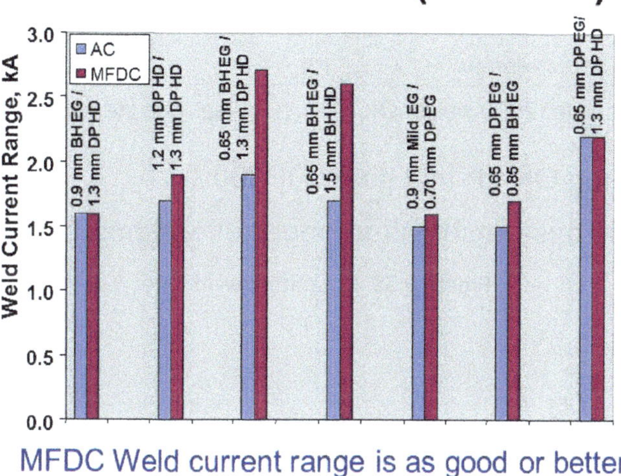

Figure 2.33 AC versus DC for IMPACT cases

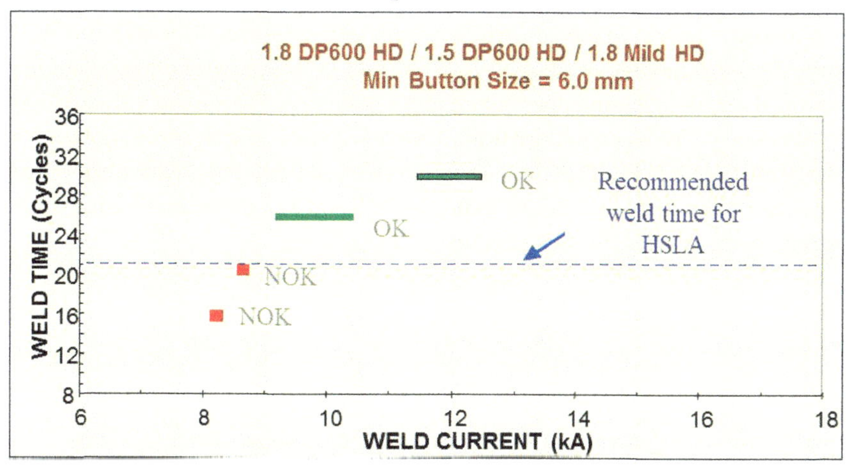

Figure 2.34 Three-thickness welding

DP500 EG / Mild Steel – AC Welding

TYPICAL CASE

0.65 mm DP- 0.8 mm Mild

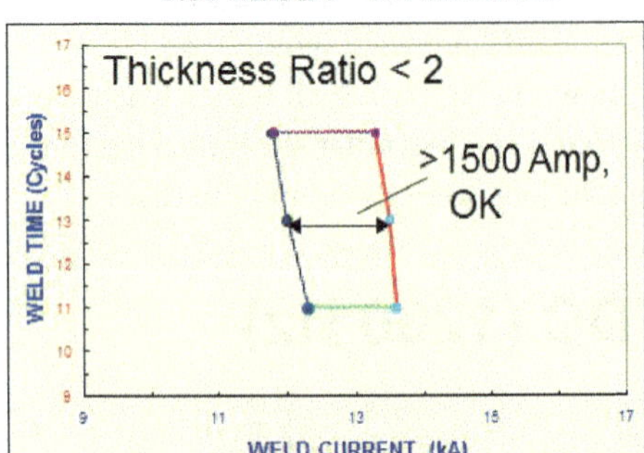

Electrode life: 2000 welds, OK

WORST CASE

0.65 mm DP- 2.0 mm Mild

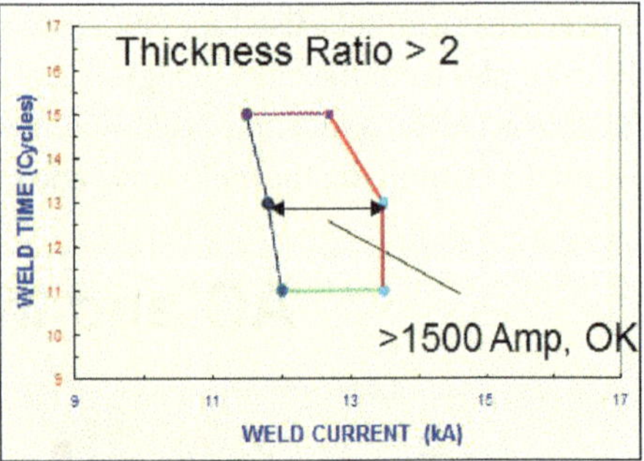

Electrode life: 1250 welds, NOK

- AC Welding OK for thin gage DP500
- Thickness mismatch has stronger effect than material/grade

Figure 2.35 Optimizing weld time

2.3 Fatigue Life with Advanced High-Strength Steels

In order to talk about fatigue, it is helpful at this point to review some fundamental fatigue theory. Figure 2.36 shows a couple of S-N curves. To create a point on these curves a tensile test specimen is cycled with a given sinusoidal load until failure occurs. The stress on the specimen and the cycles to failure are plotted as one point on the graph. This process is repeated for a large number of loads applied to different specimens until the whole S-N curve is obtained. Curve A in the figure is indicative of ferrous metals or titanium and has a flattening at a large number of cycles. Curve B would be indicative of other metals (e.g., aluminum and magnesium) and does not exhibit a flattening. The stress at which Curve A flattens out is known as the endurance limit. Theoretically, having an endurance limit means that if the stress level is below that limit, the specimen can be cycled an infinite number of times without breaking. Typically, for mild steels the ratio of the endurance limit to the ultimate tensile strength (UTS) is about 0.50.

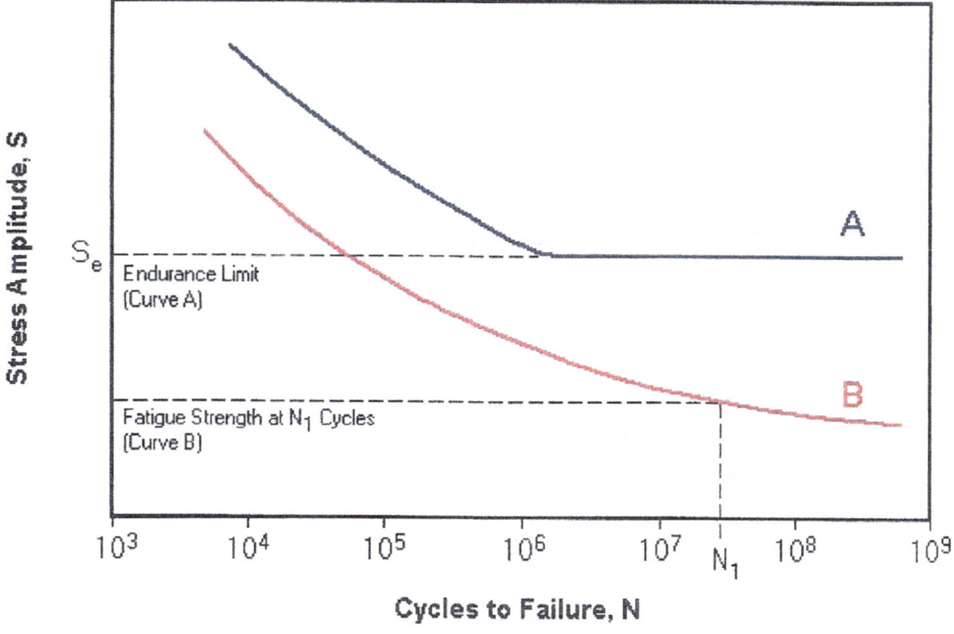

Figure 2.36 S-N curves

Fatigue performance of AHSS (Figure 2.37) is not as well understood because there has not been that much data, which is readily available for CHSS. However, some recently available data [2-7] show the following points:

1. Data for CHSS and AHSS show endurance limits to UTS ratios of about .35 to .55.
2. Compared with HSLA 350, DP 600 has an endurance limit to UTS ratio about 30% higher.
3. Compared with HSLA 350, TRIP 600 has an endurance limit to UTS ratio about 70% higher.

However, the bottom line is that all steels have endurance limits, which are roughly proportional to tensile strength. Based on this, there would seem to be a big advantage in fatigue strength in using AHSS. However, the problem is not with the steel sheet itself but with welding, because as discussed previously, the weld itself and the HAZ are closer in strength to mild steel. The fatigue strength of the weld itself, whether spot welds or GMAW welds, may only be slightly higher than welds in mild steel [2-8, 2-9]. The obvious strategy in overcoming the weld fatigue strength issues is to use more spot welds or a greater length of weld line for GMAW when using AHSS. After all, elementary design theory says that one should design so that failure occurs in the parent metal rather than at where two sheets of metal are joined. Also, with AHSS, GMAW weld joint fatigue reduction is not only caused by softer HAZ or weld nuggets. Residual stresses can reduce the fatigue limit as well. Practices used to reduce residual stresses in the weld area (such as shot peening or stress-relieving heat treatment) have been shown to be successful at increasing the fatigue limit in these welds. These practices are not currently in use for automotive structures. A very comprehensive study done on AHSS welding can be found in the study referenced at the top of the right-hand column of Figure 2.37 [2-10].

Fatigue/Durability Performance

Issues	Countermeasures *Industry Programs to address the issues*
• **Lack of fatigue data for durability analysis** • **Lack of data and knowledge for welds, including spot welds, laser welds, and MIG welds**	• **Fatigue Behavior of AHSS** – AISI-DOE TRP0038 project – Fatigue data has been generated - COMPLETED • **Sheet Steel Fatigue Taskforce** – A/SP Project - Generating spot weld fatigue data – COMPLETE IN 2003 - MIG welds – IN PROGRESS • **TWB Taskforce** – A/SP Project - Laser weld fatigue – IN PROGRESS

Figure 2.37 Fatigue performance of AHSS

2.4 Stiffness Retention When Using Advanced High-Strength Steels

Historically, one of the largest impediments to the use of AHSS steel has been the loss of stiffness as sheet metal thicknesses are typically reduced from mild steel or CHSS gauges. Since all steels have nearly equivalent moduli of elasticity, the reduced thicknesses give rise to overall stiffness reduction and resultant noise, vibration, and harshness (NVH) and vehicle handling degradation. This used to be a much larger impediment in the early 1990s when vehicle weight was primarily driven by deficiencies in vehicle stiffness. However, today this is less of a problem due to more stringent vehicle safety requirements that have appeared in the last ten years or so. This means that if a vehicle passes all the safety requirements, it will probably be sufficiently stiff for NVH and vehicle handling requirements. Also, structural improvements in vehicle architectures have reduced the functional penalties for reduced sheet metal thicknesses (Figure 2.38).

Stiffness Retention When Using AHSS

Issues	Countermeasures *Industry Programs to address the issues*
• Replacing conventional steels with thinner gage AHSS could result in a loss in stiffness	• Improvements made to vehicle architecture could counteract the stiffness losses – some examples later in the seminar • CAE could be utilized to better optimize gages across the vehicle to recover stiffness loss. • CAE could be utilized to optimize the location of structural adhesives to provide extra stiffness.

Figure 2.38 Stiffness retention

One strategy being used today for preventing a loss of stiffness with AHSS utilization is CAE methods for gauge optimization. Methods that were being used in the 1990s focused on using linear vehicle models for NVH optimization [2-11]. The method discussed in this paper involved using FEA models of the major vehicle subsystems, developed using NASTRAN; combining these models in a system model using a program like MOTRAN; calculating the sensitivity to gauge reduction using a program such as CONTESA; and then optimizing the whole vehicle using an optimizer such as SYSOPT to minimize frequency response amplitudes (Figure 2.39). Instead of optimizing for durability, safety, and vehicle handling directly, these attributes were addressed as constraints to the optimization analysis.

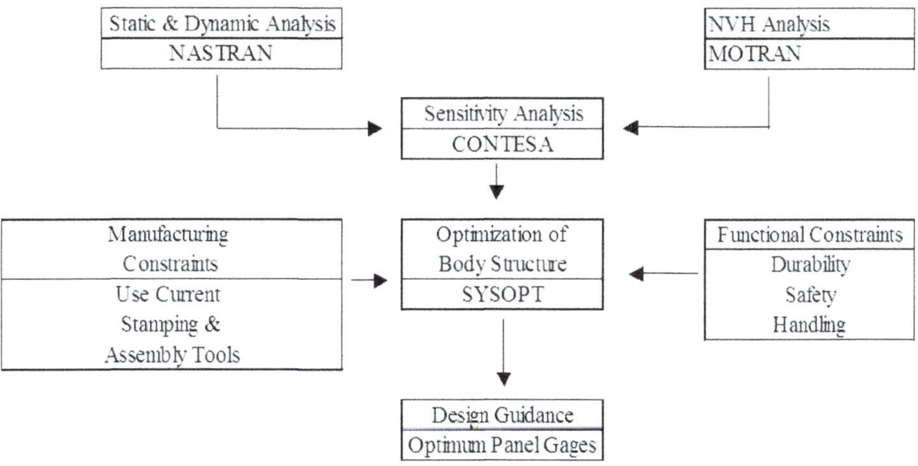

Figure 2.39 Thickness optimization utilized in the early 1990s

More commonly today, the thickness optimization is performed in two phases [2-12]. First, it is not usually necessary to create a vehicle system model. A body structures model done in NASTRAN is sufficient for the NVH and vehicle dynamics optimization. Certain body stiffness and modal frequency metrics are usually known today to provide appropriate NVH and vehicle dynamics. An optimizer program such as Genesis is used in conjunction with the NASTRAN FEA model to produce the minimum weight by reducing the sheet metal thicknesses. However, in some cases the optimizer will increase thickness of some parts to achieve compensating thickness reductions in other parts, for a net weight decrease. Then, a nonlinear FEA (e.g., LS-DYNA) model of the body structure and the linear stiffness derived gauges from the previous analysis is used in conjunction with an optimizer such as HEEDS to minimize thicknesses, while achieving appropriate levels of critical crash responses. Because of the dominance of vehicle safety in driving vehicle weight today, rerunning the NASTRAN NVH runs using these gauges typically shows an improvement in NVH performance without having to rerun the NVH optimization.

Another method of recovering lost stiffness with AHSS usage is by using structural adhesives or continuous laser welding. The structural elements of the body are typically formed by "shingling" two pieces (or more) of sheet metal together and spot welding to form a beam. Spot welds are placed some nominal distance apart (50 mm), which means that in between the spot welds the beam can actually act like an open section beam with significantly less stiffness than for the sections immediately within the spot welded nugget. Bodies that use structural adhesives or laser welding can be as much as 20% stiffer than bodies without these enhancements. The good news is that you only have to use laser welding or structural adhesives on about 20% of the seams in the body to get about 90% of the benefit. Because of this, methods have been devised for optimizing the location of structural adhesives or laser welds [2-13].

2.5 Computer-Aided Engineering and Advanced High-Strength Steels

We have already mentioned stiffness retention using CAE optimization methods. There are a few other CAE related issues that need to be considered to fully benefit from AHSS usage (Figure 2.40). First, good AHSS properties need to be obtained for good AHSS calculations. We have already mentioned the lack of good forming limit curves for some AHSS. The lack of good forming limit data may necessitate using soft tools for some parts to validate the forming analysis at the same time that the industry is trying to get away from having to generate soft tools by using CAE. High strain rate data also need to be obtained. Stresses will increase at any given value of strain, depending on how high the strain rate is. This data is readily available for most mild steels and CHSS but is typically not available for all AHSS. This problem is complicated by the fact that very few labs, outside of the steel industry, have the capability of creating accurate strain rate sensitivity data.

Risk Assessment — CAE

- **Need good properties for accurate predictions:**
  - ⇒ **Formability Analysis — forming limit diagram**
  - ⇒ **Crashworthiness — high strain rate data, forming hardening, thinning, BH**
  - ⇒ **Durability — fatigue data for base metal and spot welds**
- **Quickly validate CAE for CAD=>hard tool direct transition**
  - ⇒ **May need to go back to soft tools in some cases**
- **Acceptable rules must be published**

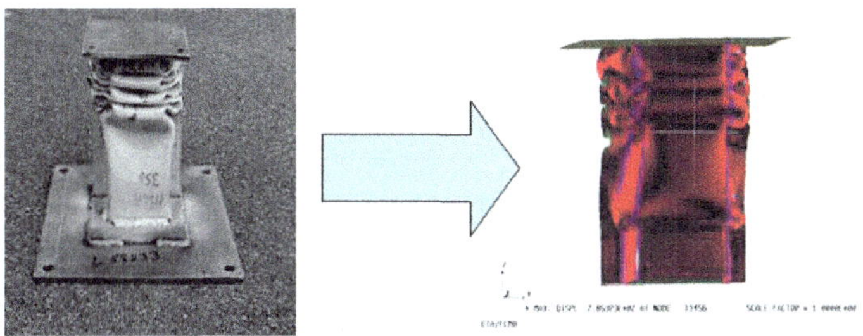

Figure 2.40 CAE risk assessment

Since many AHSS have fairly high tensile to yield strength ratios, the "after forming" stress-strain curves can be significantly different than the preforming properties. Also, forming tends to thin the steel, which also may have to be considered. The only way to accurately do this is to do an incremental forming analysis using LS-DYNA or similar FEA code and map the properties and the thickness characteristics into the FEA bulk data for subsequent attribute analysis. However, this is very tedious and will add considerable analysis time to the product development cycle. Investigations are currently ongoing at the steel companies, the OEMs, the Tier 1s, and the A/SP to figure out an economical way of performing this analysis [2-14]. The good news here is that increases in stress data are at least partially offset by thinning; and simple beam sections that are primarily bent versus drawn and that are subjected to longitudinal loading may not be influenced that much by changes during forming. A related issue is for AHSS that have a significant amount of bake hardening (BH) during the paint cycle should be represented by appropriately elevated stress-strain data.

In summary, the major issues for CAE in successful implementation of AHSS are a lack of high strain data and the lack of an FEA guideline for crash simulation (Figure 2.41). There are several major studies ongoing to address these issues, shown in the right-hand panel of Figure 2.41. One major study [2-10] addresses strain rate sensitivity data.

Role of CAE in Successful Implementation

Issues	Countermeasures *Industry Programs to address the issues*
• **Lack of high strain rate data for AHSS for crash analysis** • **Lack of FEA guideline for crash simulation**	• **High Strain Rate Behavior of AHSS** – AISI-DOE TRP0038 project – generated high strain rate tensile data for AHSS - COMPLETED • **Developing Recommended Practice for High Strain Rate Testing** IISI-A/SP project • **Developing FEA Guidelines for Crash Analysis** – A/SP project • **High-Strength Multiphase Sheet Steels under Dynamic Deformation Conditions** – AISI Project 9904

Figure 2.41 Successful implementation using CAE

2.6 Variability of Advanced High-Strength Steels

Variability of AHSS has been identified as a major impediment for their utilization. However, some data suggest that within a given mill, the variability of AHSS is no different from CHSS. Figure 2.42 shows some data from a given mill where DP 500, an AHSS, is compared to three CHSSs. In terms of the probability density functions, shown at the bottom of this figure, the DP 500 is no worse or slightly better in variability of significant mechanical properties. However, it is well known that there is much larger mill-to-mill variability of AHSS versus CHSS. One reason for this is that some mills use more chemistry to create AHSS properties, and some mills use more process in terms of heating and cooling to create AHSS. One of the examples, reviewed later in this book (A/SP Closure Study), will demonstrate some of the variability in steel competitors' AHSS properties. There is some considerable effort ongoing at the individual auto companies and at the A/SP to create an expanded specification list, which will push the steel companies to produce more uniform properties for a given AHSS grade.

Accommodating AHSS Property Variability

Some Considerations:

1. **Industry is still in development phase – in some cases, each batch is of a refined recipe (coil-coil variability)**
2. **Patents, unique mill capabilities necessitate differing approaches to difficult ends (vendor-vendor variability)**
3. **Some Mills claim DP500 data (~20 coils) less variable than HSLA 350 – will improve with experience:**

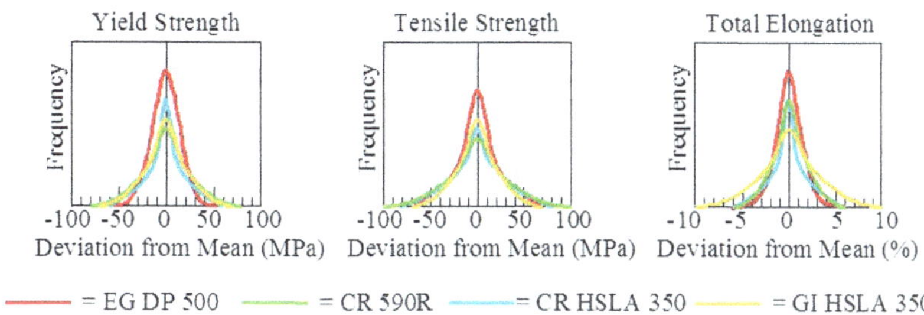

Figure 2.42 AHSS variability

2.7 Service and Repair of Advanced High-Strength Steels

Another area that is a major concern with any application of AHSS is service and repair (Figure 2.43). To understand this issue, it is important to understand that auto companies and insurance companies want to minimize cost and therefore avoid scrapping out a vehicle if at all possible. The repair of vehicles should be done in such a way that the after-crash structural integrity should be equivalent to the pre-crash structural integrity. That being said, permanent deformation after a crash or other major event was handled historically by heating the deformed member(s) and bending the parts back into the pre-crash state. However, as is shown in Figure 2.43, that procedure could be a disaster with AHSS, since annealing the metal and cooling can create significantly different mechanical properties from the original design. For this reason, procedures have to be developed for the common accident scenarios to cut out deformed members, replace them with new parts, and GMAW (mig) them into position. The developed procedure may have to be validated with a crash test. Though this procedure adds considerable difficulty to the product development cycle, it is probably not as costly and time consuming as that which would be required with aluminum intensive vehicles (AIVs).

Risk Assessment – Service/Repair

Effect of flame straightening exposure on GI DP 600 strength

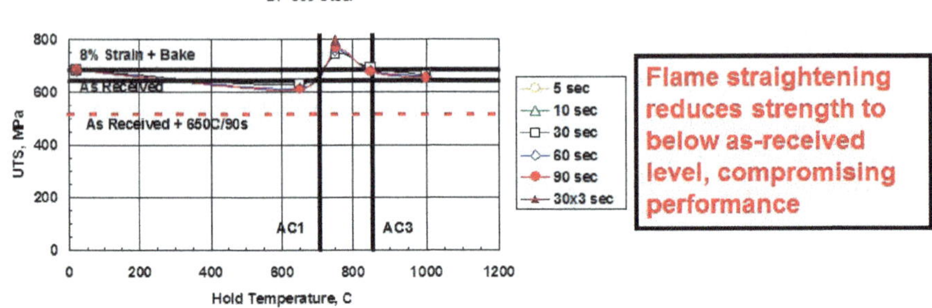

- **Avoid application of heat during repair**
- **Cut-and-MIG weld = preferred option**
- **Repair procedures should be verified with crash test**
- **Still not as difficult to repair as AIV!**

Figure 2.43 Service/repair risks with AHSS

2.8 References

2-1. Sadagopan, S., Urban, D., 2003, *Formability Characterization of a New Generation of High Strength Steels,* U.S. Department of Energy No. DE-FC07-97ID13554, Washington, DC.

2-2. Auto/Steel Partnership (A/SP) AHSS Applications Guidelines Group, 2010, *Advanced High-Strength Steel Application, Design and Stamping Process Guidelines.* Auto/Steel Partnership Publication, Southfield, Michigan, link available at www.a-sp.org, accessed in January 2013.

2-3. Gericke, Dean, 2012, "Hydroform Intensive Body Structure with Advanced High Strength Steels," presentation to the 2012 Great Designs in Steel Conference, Livonia, Michigan, May 2012, available at www.autosteel.org, accessed in January 2013.

2-4. A/SP, 2000, *High Strength Steel Design Manual,* Auto/Steel Partnership Publication, Southfield, Michigan.

2-5. Young, Darryl, 2009, "Investigation of Tooling Durability for Advanced High-Strength Steel," presentation to the 2009 Great Designs in Steel Conference, Livonia, Michigan, available at www.autosteel.org, accessed December 2012.

2-6. A/SP, 2007, *An Investigation of Resistance Welding Performance of Advanced High-Strength Steels,* Auto/Steel Partnership Publication, Southfield, Michigan.

2-7. Yan, B., 2003, "Fatigue Behavior of Advanced High Strength Steels for Automotive Applications," presentation to the 2003 Great Designs in Steel Conference, Livonia, Michigan, available at www.autosteel.org, accessed February 2013.

2-8. Chiang, J., Jiang, C., 2005, "Fatigue Performance of DP600 Resistance Spot Welds made with Production-Friendly Welding Schedule," presentation to the 2005 Great Designs in Steel Conference, Livonia, Michigan, May 2005, available at www.autosteel.org, accessed January 20013.

2-9. Anderson, D., Hunt, J., Sang, J., 2009, "Effects of Material Properties and Weld Geometry on Fatigue Performance of DP780 and Mild Steel GMAW Lap Joints," Presentation to the 2009 Great Designs in Steel Conference, Livonia, Michigan, May 2009, available at www.autosteel.org, accessed February 20013.

2-10. Yan, B., Urban, D., 2003, *Characterization of Fatigue and Crash Performance of New Generation High Strength Steels for Automotive Applications (Phase I and Phase II),* U.S. Department of Energy, Work Performed under Cooperative Agreement TRP0038, prepared by AISI Technology Roadmap Office, Pittsburgh, PA 15222.

2-11. Choi, Kyung K., Kulkarni, Harihar T., 1991, "Design Sensitivity Analysis of Dynamic Frequency Responses of Acousto-Elastic Built-Up Structures," Proceedings of the NATO, DFG Advanced Study Institute on Optimization of Large Structure Systems, Berchtesgaden, Germany, Sept. 23–Oct. 4, 1991.

2-12. EDAG Inc., 2012, *Mass Reduction for Light-Duty Vehicles for Model Years 2017–2025,* DOT HS 811 666, U.S. Department of Transportation, National Highway Traffic Safety Administration.

2-13. Jasuja, Subhash C., Kulkarni, Harihar T., Geck, Paul E., 2004, "System and Method for Designing Automotive Structure using Adhesives," U.S. Patent Office, Patent No. 6766206.

2-14. Sohmshetty, R., Ramachandra, R., Bhimaraddi, A., 2010, "Forming Effects to Product Attribute Coupled CAE Process and Benefits Investigation," SAE Paper No. 2010-01-0448, SAE International: Warrendale, PA.

Chapter 3
Example Applications of Advanced High-Strength Steels

3.1 Architectural Enablers for Advanced High-Strength Steels

This chapter will serve as a study of various applications of advanced high-strength steels (AHSS), which are meant to convey some significant AHSS concepts. However, first it will be instructive to review various architectural enablers that automotive manufacturers have been using to enable or enhance the weight reduction effects of AHSS. This section will identify these enabling technologies and explain the engineering and business trade-offs and will identify what vehicle parts the enabling technologies can be applied to. The rest of the chapter will focus on applications that use the various technologies.

The first enabling technology is laser-welded blanks (LWBs). With this technology, two sheets of steel are laser-welded together (Figure 3.1) and the resulting blank is used in the press as a single sheet. Of course, the die, which is used to form the part, has to have a step to accommodate the blank if two different gauges are used. In this technology, two or more different gauges are used and/or two or more different steel grades can be used. This technology actually pre-dated widespread application of AHSS, because there are sometimes advantages for the use of this technology with mild steel or conventional high-strength steels (CHSS).

Figure 3.1 Laser welded blank

The most common application of laser-welded blanks is inner doors. Figure 3.2 shows various configurations of laser-welded blanks and patch blanks, which were studied as part of the Auto/Steel Partnership (A/SP) Door Project (discussed in detail later in this chapter). Patch blanks are a little different than laser-welded blanks in that there is some overlap of the two sheets and they are spot-welded together instead of laser welded. The center picture of Figure 3.2 is the baseline laser-welded blank configuration. Typically, for door-inner applications of laser-welded blanks, the steels are mild steels; the door-inner tends to be stiffness driven, which gives no advantage to using AHSSs because all steels have essentially the same modulus of elasticity.

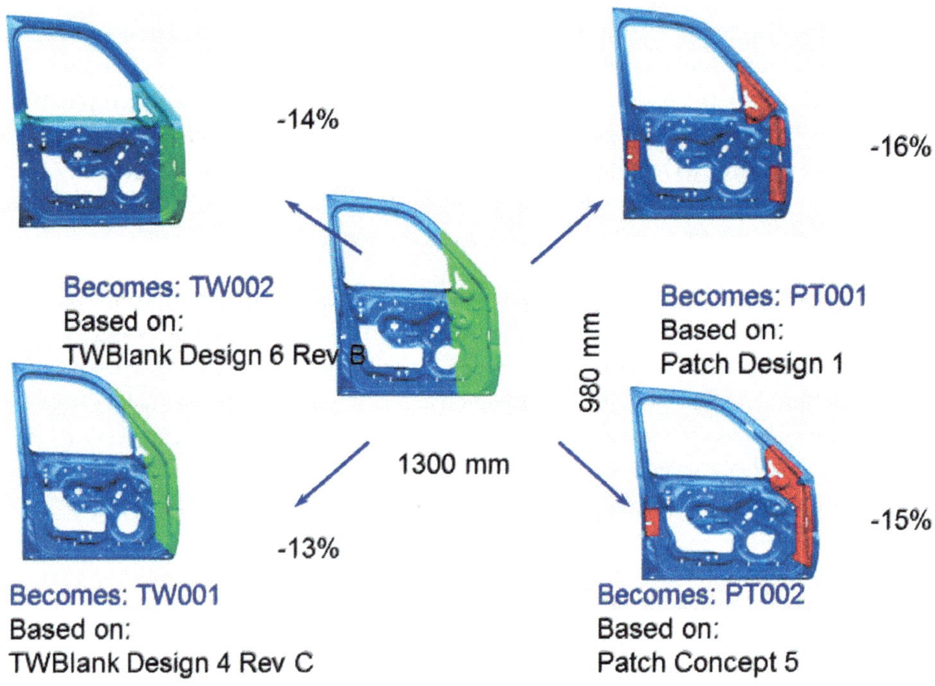

Figure 3.2 Door-inner applications of laser-welded blanks

More recently, manufacturers have been using laser-welded blanks for a large variety of structural members. For instance, front rails have been made from laser-welded blanks (Figure 3.3) using different AHSS grades and thicknesses to achieve a controlled crush and to maintain the integrity of the passenger compartment during frontal impacts.

Rail Applications

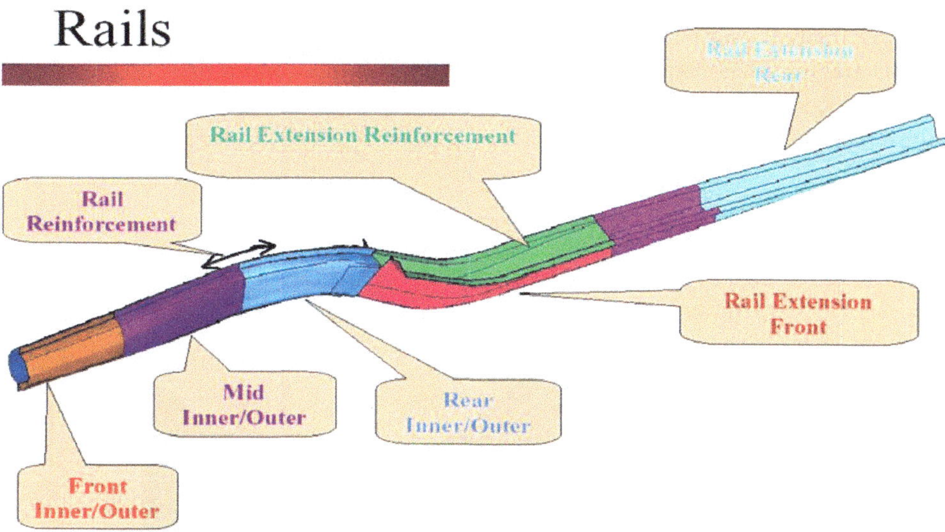

Figure 3.3 Example of laser-welded blank for front rail application

The following are advantages of using laser-welded blanks:

1. The ability to vary the thickness and steel grade over the length of a beam or panel leads to better weight optimization.
2. The investment for both tools and facilities is essentially the same as with conventional stamping.
3. For panels and structural members with reinforcements, replacement with a laser-welded blank can provide piece cost and subassembly cost reductions.
4. Laser-welded blanks can typically be made in AHSS.
5. Use of laser-welded blanks can improve blank nesting, which will increase material utilization, and which can then decrease costs.

The following are disadvantages or limitations of using laser-welded blanks:

1. Typically, laser-welded blanks with over a 2-to-1 thickness ratio are not used, but in some cases up to a 3-to-1 variation has been used.
2. The cost of the blank, estimated at about $.08 per linear inch of weld line, constrains the usage to parts with short weld lines when compared with the size of the part.
3. Parts that are formed with tight radii or deep draws in the neighborhood of the laser weld may have a problem with splits.

Another architectural enhancement mechanism, which may enhance the benefits of AHSS, is tube hydroform construction. Essentially hydroform construction takes bent tube construction one step further, where a bent tube can be put into a press in which high-pressure water is shot into the tube while it is being pressed. This can enable the bent tube to vary in diameter and thickness along the length of the tube. An example of a hydroform front end for a vehicle is shown in Figure 3.4. This construction has become popular in roof rails and pickup truck front ends. Tube hydroforms have been made out of DP steels up to DP 980.

Hydroform Front End

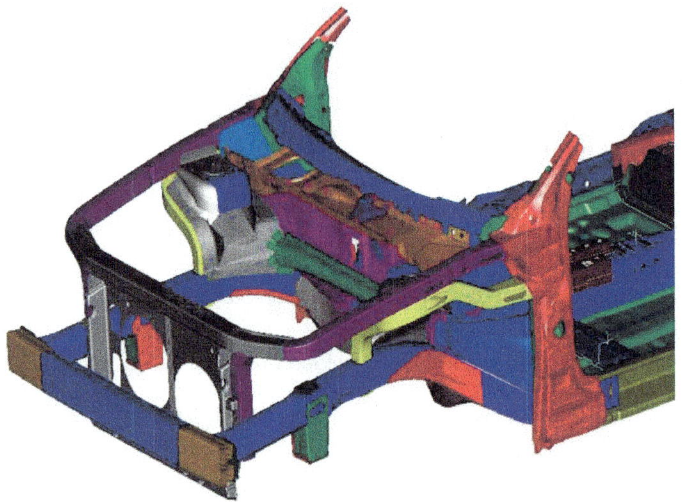

Figure 3.4 Tube hydroform front end

Tube hydroforms have found usage in a wide variety of industries, including the home appliance industry. The door handles of the GE Profile kitchen appliances (Figure 3.5) have

Door Handles

Figure 3.5 GE refrigerator door handles

been done as tube hydroforms. Note the flared ends of the refrigerator door handles, which are made out of stainless steel. Another type of hydroforming, which has found some limited application in the automotive world, is sheet hydroforming, in which a one-sided die is used in conjunction with a bladder with water under pressure backing the bladder. Because sheet hydroforming distributes stresses more evenly, there exists the possibility of using higher strength steels than with conventional stamping. There have also been claims that sheet hydroforming will decrease tooling to roughly one half of conventional stamping because of the one-sided die. This claim hasn't been fully realized because of extra stamping operations, which are often required. In any event, tube hydroforming has been a very high growth area over the last several years, where sheet hydroforming has been used primarily on low-volume specialty cars.

There are several advantages with using tube hydroform construction:

1. Tube hydroforming can provide constant or varying section properties over the length of the beam.
2. Tubular sections provide good stiffness and strength for multidirectional loading.
3. Tubular hydroform, space frame subsystems are ideal for modular construction applications.
4. Tube hydroforms can provide more design flexibility versus simple bent tube construction.

Some of the disadvantages of using tube hydroforms follow:

1. It is sometimes difficult to integrate with sheet metal construction from a welding perspective.
2. Space frame tube construction does not necessarily provide the lightest weight construction when compared with a well-integrated monocoque construction.
3. It was thought that tube hydroforms were limited to medium-strength AHSS, but more recently the use of DP 980 on the Ford Fusion shows that this limitation may have been exaggerated.

Another popular enabling technology being used in conjunction with AHSS is roll forming, where a part is formed by passing the sheet metal blank through a series of rollers, which gradually form the part by bending the sheet metal a little bit more with every new set of rollers. Roll forming is a fairly old technology, which is reinventing itself because of the use of high-strength steels, where roll forming can even be used even with ultra-high-strength martensitic steels. Some examples of roll-formed parts are shown in Figure 3.6. Frame midrail sections are commonly used with full-frame vehicles, where strength can help in side impact situations. Many of the parts of the GM pickup truck beds have historically been roll formed. Roll forming can be used to create complicated multicell sections, as shown in the roof rail in the figure. Another recent example is the use of roll-formed rocker sections, illustrated in Figure 3.7.

Roll form Design Opportunities

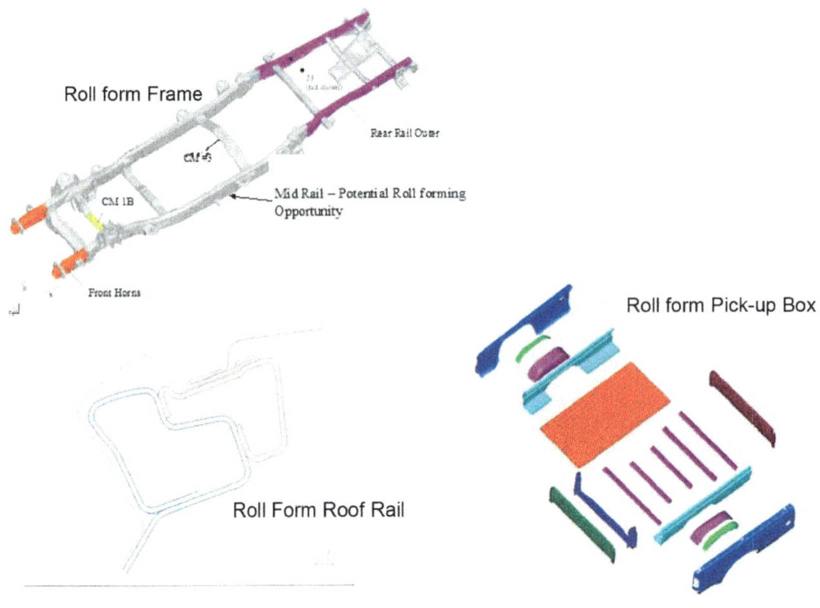

Figure 3.6 Roll form design examples

Body Side Lower / Rocker Concept

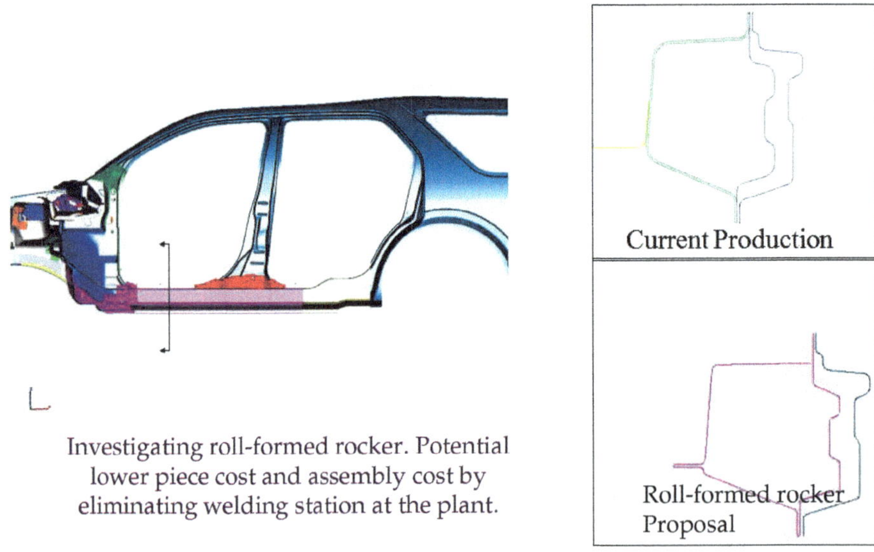

Figure 3.7 Roll form rocker

The following are advantages of roll forming:

1. Much higher strength steels can be roll formed than with traditional stamping.
2. Investment for both tools and facilities can be less than with traditional stamping, especially if new presses are required for a particular high-strength steel stamping operation.
3. Cost and setup of roll stands usually only need to be considered for initial production. The rolls themselves may need to be modified for a model changeover, but not the roll stands.
4. Variable costs can typically be lower than with conventional stamping, especially if thinner gauges are used.
5. Joining roll-formed components to sheet metal components is usually easier than with tubular components because it is easy to roll form weld flanges, and martensitic steels are quite weldable.

The following are disadvantages of roll forming:

1. Often it is hard to integrate roll-formed parts with conventional sheet metal construction because roll-formed parts typically have uniform, straight sections that may require architectural changes to the mating parts.
2. Post-operations are necessary if the section of the part varies over the length or if complex bends are required.
3. Cycle time for a roll-formed part can be longer than with conventional stamping.

Another architectural enhancing technology is Tailor rolled blanks (TRBs), which are marketed by Mubea Inc. In this technology, a coil of steel is uncoiled and sent through rollers that are able to move vertically on their axes, which can impart a changing thickness over the length of the coil (Figure 3.8). Figure 3.9 shows some potential applications for TRB.

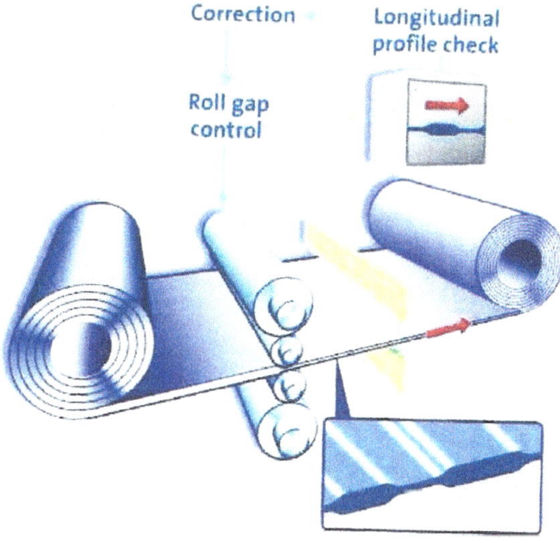

Figure 3.8 Tailor rolled blanking

Probably the most impressive application of TRB is the major B-pillar reinforcement on the Ford Focus, which is the primary protection for side impacts. This B-pillar reinforcement has eight separate thicknesses progressing vertically up the B-pillar (Figure 3.10). To do this application with laser-welded blanks would be very expensive.

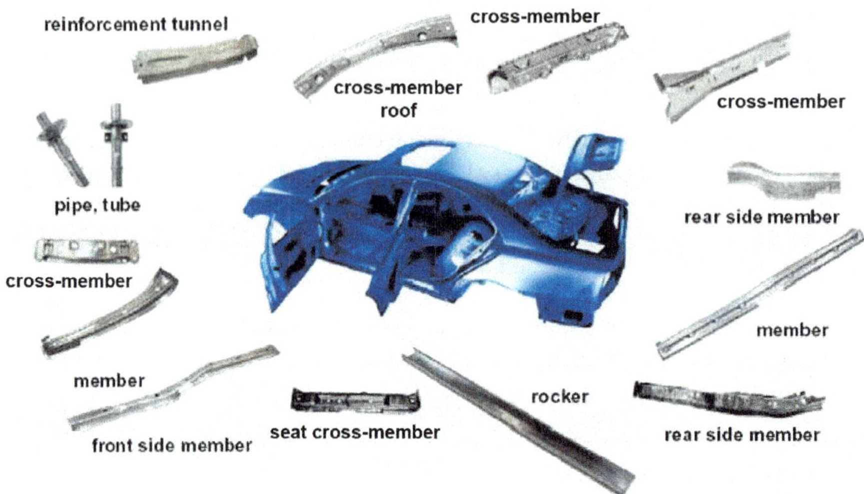

Figure 3.9 Possible Tailor rolled blank applications

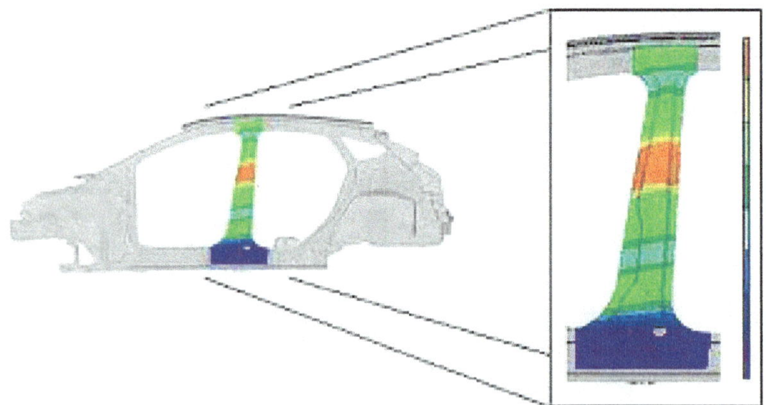

Figure 3.10 Ford Focus Tailor rolled blank B-pillar

The following are advantages of TRBs:

1. The thickness can vary over the length of the beam.
2. For the original equipment manufacturer (OEM) using the TRB, the investment for both tools and facilities is approximately the same as with conventional stamping.
3. For narrow structural members with reinforcements, replacement with TRBs can provide piece cost and subassembly cost reductions.
4. Laser-welded blank structural members, with several blanking lines, can be replaced with TRBs for a cost advantage.
5. For large parts and panels where a continuous thickness change is required, TRBs can provide a cost advantage.

The primary disadvantage for TRBs is that the strength of the steel, the size of the part, and the thickness of the various sections of the sheet will dictate the power required to drive the reduction rolls. Power limitations have been a major impediment for the use of TRBs. Another disadvantage of using TRBs is the need to orient the part consistent with the orientation of the gauge transitions. This makes orienting the part relative to the coil (for improving blank nesting) almost impossible for most applications, and TRB parts generally suffer from lower material utilization (higher scrap cost) than in the case of LWBs. Generally, the TRB approach will only be cost effective if the blank is quite rectangular and easy to nest. Otherwise, the better scrap utilization of the LWB approach will dominate. However, each application needs to consider all aspects to make the proper, lowest cost solution.

The fastest growing enabler of AHSS, in terms of the number of parts stamped, is hot stamping. This technology has been used for many years in simple parts such as internal side intrusion door beams, where extremely high strength is required. The steel, which is used for this process, is called hot-stamped boron steel or press-hardened steel (PHS). We will use these two designations interchangeably in this text. The ever-increasing safety and fuel economy requirements have meant that car companies are using hot-stamped beams for complex structural parts, where stamping is prohibitive with higher strength steels. For instance, roof components and side impact components such as the primary B-pillar reinforcement are now being done by some manufacturers using hot stamping. Hot stamping is accomplished after blanks are cut by heating hot stamping steels to temperatures above the austenitic temperature (Figure 3.11). The blank is pressed in a special die, which is cooled internally with water-cooled channels, and held in the press until cool. The part is then shot blasted to remove the resulting oxide scale, which coats the part. The need for shot blasting is eliminated if an aluminum coating is put on the steel before blanking. Aluminized boron steel is now commonly used to displace bare boron steel.

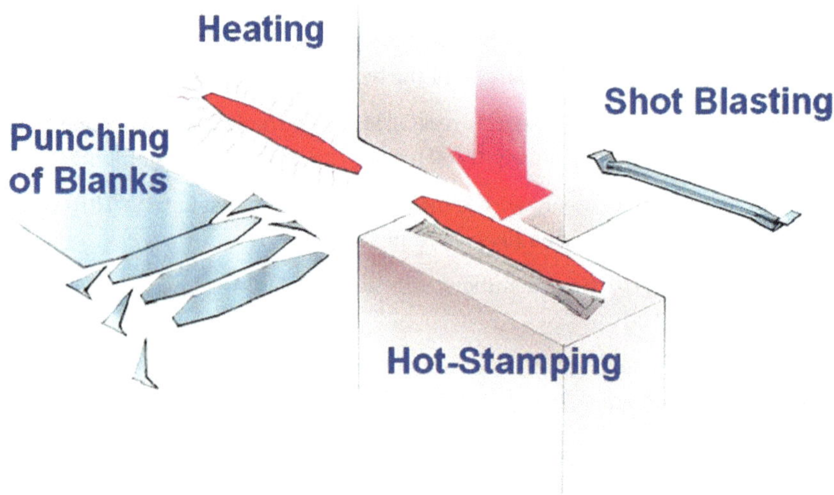

Figure 3.11 Hot stamping process

There are some significant advantages for the use of hot stamping:

1. A very high strength (1500 MPa tensile strength) can be produced in otherwise hard-to-stamp geometries. For instance, springback is dramatically reduced.
2. It can provide a very low weight solution for crash applications.
3. Welding can be easier than with ultra-high-strength, multiphase steels.
4. Current architectures do not have to vary to integrate a hot-stamped part.
5. The costs of uncoated boron can be somewhat less than with some of the common high-strength AHSS.
6. Part springback is significantly reduced compared to stamped AHSS parts.

However, there are some disadvantages that need to be considered:

1. Hot-stamped facilities cost a lot.
2. Thermal cycling results in die fatigue, resulting in increased die maintenance costs.
3. Extreme strength of parts can present repairability problems.
4. Low ductility may lead to premature cracking in some high-deformation situations.
5. Cycle time is slowed because the part has to be heated and held in the press until it is cooled. This disadvantage is being addressed by having parallel furnaces and presses.
6. Probably the biggest disadvantage of hot stamping is the effective variable cost. If the extra processing costs are added to the cost of the material, the net variable cost can be about double the cost of other AHSS and only slightly less than austenitic stainless steel, which is actually a second generation AHSS [3-1].

However, the higher costs associated with hot stamping are often mitigated by a dramatic reduction in thickness or the elimination of reinforcements. The total system costs of a hot-stamping application must be considered to determine the cost impact.

To compare these architectural enablers, we need to address the advantages and disadvantages of conventional stamping operations. With conventional stamping, parts that require a different material or gauge are stamped separately and then are welded together (Figure 3.12).

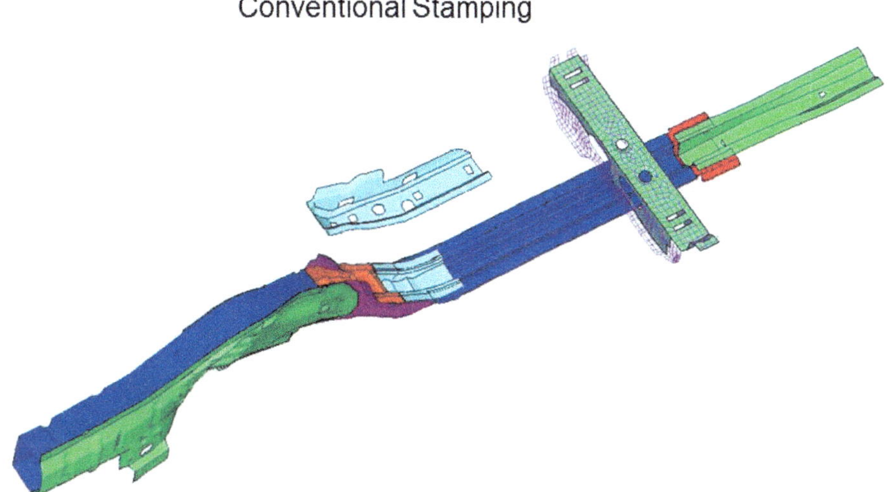

Figure 3.12 Front rail subassembly made with conventionally stamped parts

When compared with the enablers that we have been discussing, there are some advantages with conventional stamping:

1. Conventional stamping facilitates monocoque construction, which can provide cost advantages for high-volume construction versus space frame architectures.
2. Spot welding typically is easier with conventionally stamped parts.
3. It is easy to add reinforcement pieces to give localized strength where it is needed.
4. Stamped parts can utilize existing stamping presses, which helps reduce upfront investment.

However, there are a couple of disadvantages:

1. Though there are some recent examples of conventionally stamped DP 980 and martensitic steels, conventional stamping is typically limited to DP 780, in terms of strength.
2. Without architectural enablers, one will typically end up with the greatest number of parts.

The table in Figure 3.13 compares the major enablers or fabrication methods. The table was constructed primarily from an OEM perspective. However, this table is highly subjective and a comparison should really be done with a specific part in mind.

Fabrication Method Comparison

	Hydroforming	Rollforming	Tailor Rolled Blanks	Tailor Welded Blanks	Hot Stamping	Conventional Stamping
Applications	Structural members	Crossmember& Corrugated/ Flat panels	Crossmember	Crossmember & panels	Structural members	Any
Enabler for AHSS	Limited to 980 MPA	Yes	Limited to 350MPA	Limited to 980 MPA	When UHSS Is needed	Limited to 780 MPA
Joining Difficulty	High	Medium	Low	Low	Low	Low
AHSS Forming Difficulty	High	Low	High	High	Low	High
Assembly Difficulty	High	High	Low	Low	Medium	Low
Investment	Medium	Medium	Low	Low	High	Low
Piece Cost	High	Low	High	High	High	Medium

Figure 3.13 Comparison of major architectural enablers or fabrication methods

In summary, it might be said that several competing architectural enablers are being used to facilitate or enhance the use of AHSS. A few of the enablers (e.g., roll forming, laser-welded blanks, and hot stamping) are already being used to a fairly high degree, with hot stamping being on a fast high-growth curve.

3.2 IMPACT Applications

The next two sections will focus on applications. The applications have been selected based on the learning potential for understanding these applications and on the author's understanding of the applications, in that the author had a major role in the performance of these applications. The first set of examples is from the Improved Materials and Powertrain Concepts for 21st Century Trucks (IMPACT) project (Figure 3.14). This was a major cross-organizational study, funded by Ford Motor Co., the U.S. army, and the American Iron and Steel Institute. Also, the University of Louisville and Mississippi State University had major roles in the project. The overall objective of this project was to design and validate fuel efficient, lightweight technologies for next generation, high-volume, commercially based truck platforms. IMPACT was performed in three phases (Figure 3.15). Most of this section will focus on IMPACT Phase II, which was a study to remove 25% of the weight from a full-size pickup truck. There are also a few examples from IMPACT Phase III, which was a weight reduction study for super-duty pickup trucks. IMPACT Phase I was an alternative materials study, which will be covered in part in the next chapter. Though it was a research study, the significance of the IMPACT study can be measured in that over 60% of the technologies, which were developed, were adopted on Ford vehicles.

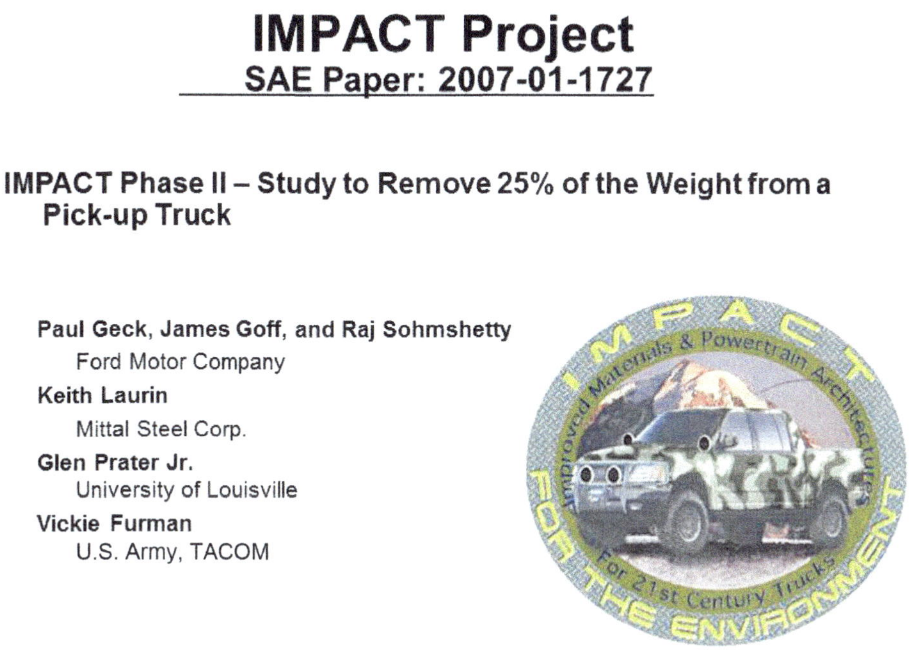

IMPACT Project

SAE Paper: 2007-01-1727

IMPACT Phase II – Study to Remove 25% of the Weight from a Pick-up Truck

Paul Geck, James Goff, and Raj Sohmshetty
Ford Motor Company
Keith Laurin
Mittal Steel Corp.
Glen Prater Jr.
University of Louisville
Vickie Furman
U.S. Army, TACOM

IMPACT
Improved Materials & Powertrain Architectures
For 21st Century Trucks
FOR THE ENVIRONMENT

Figure 3.14 IMPACT Project

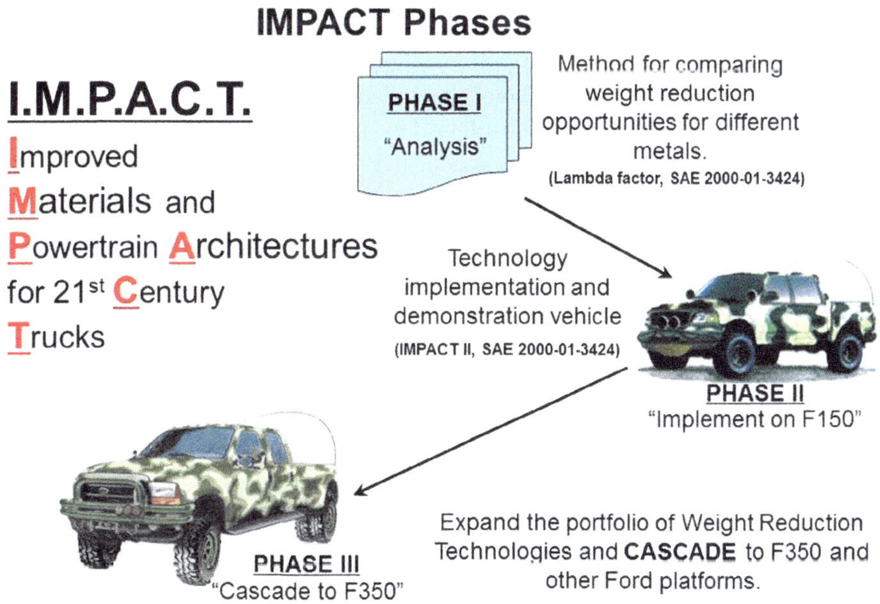

IMPACT Phases

I.M.P.A.C.T.
Improved
Materials and
Powertrain Architectures
for 21st Century
Trucks

PHASE I
"Analysis"

Method for comparing weight reduction opportunities for different metals.
(Lambda factor, SAE 2000-01-3424)

Technology implementation and demonstration vehicle
(IMPACT II, SAE 2000-01-3424)

PHASE II
"Implement on F150"

PHASE III
"Cascade to F350"

Expand the portfolio of Weight Reduction Technologies and **CASCADE** to F350 and other Ford platforms.

Figure 3.15 IMPACT phases

When developing IMPACT Phase II, several competing technologies were selected for each subsystem. The technologies were evaluated using the following criteria, to enable the team to determine the winning technology:

1. Implementable on Ford vehicles between model years 2004 to 2007.
2. Each subsystem was expected to replace the current model with a 25% weight reduction subsystem. In managing the whole project, account was taken of the fact that some subsystems could easily achieve the target whereas other subsystems could not exactly meet the target.
3. The IMPACT vehicle was designed with carryover vehicle attributes (e.g., safety, NVH, durability, and so forth).
4. Steel as a material was not a requirement, but if there were two closely competing technologies, the steel version was chosen.
5. The chosen technologies had to be implementable in a demonstration vehicle by the project material ready date.
6. The competing subsystems were evaluated on variable cost and weight saves. There was no investment constraint.
7. All the selected subsystems had to support fuel economy improvement. In other words, if a technology met the weight target but increased fuel usage (e.g., increased wind resistance), it was not selected.
8. In some cases, the teams were challenged to improve performance while reducing weight.
9. System level degradation was allowed if vehicle performance was unaffected.

An example of the last point could be found in the frame system stiffness target. One of the top drivers for pickup frame stiffness is minimization of low-frequency freeway hop caused by road inputs. In the IMPACT project, computer-aided engineering (CAE) revealed that the frame stiffness could be lowered and still lead to an improvement in freeway hop if the mode shape of the frame bending mode was adjusted such that the rear nodal point of the cab-box-frame combination was close to the rear suspension input points (Figure 3.16).

In order to simultaneously reduce weight and cost, technologies were sorted using a weight value curve (Figure 3.17). To generate this curve, the highest value (weight reduction divided by cost increase) alternative for each subsystem was determined. All the selected subsystems were rank ordered, with the highest value subsystems being plotted at the left side of the graph. The full plot was then constructed with variable cost decreases plotted against cumulative weight reduction. Note that many of the technologies, up to about 500 lb weight reduction actually reduced cost or were cost neutral, but beyond this point cost was actually being added. The target of 25% or 1310 lb weight reduction was met with only about $800 of added cost.

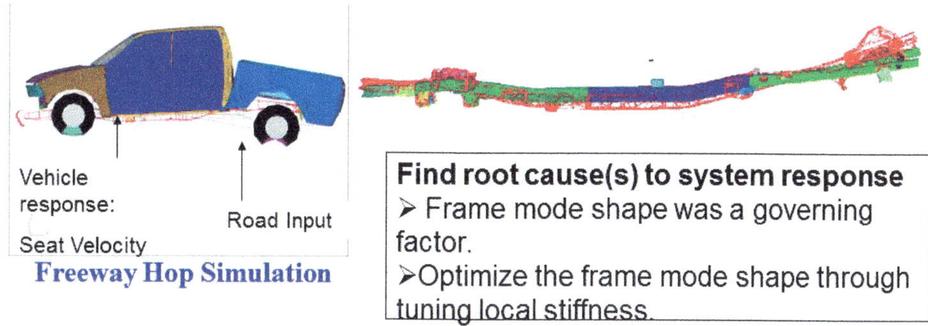

Figure 3.16 Freeway hop simulation

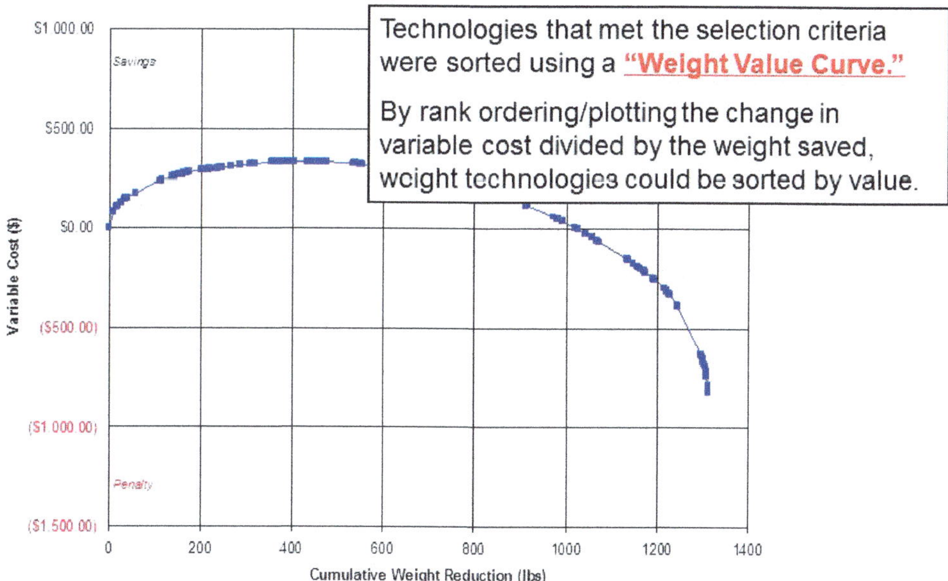

Figure 3.17 IMPACT value curve

The material focus for IMPACT Phase II was to convert mild steel to CHSS or the lower end of the AHSS range (Figure 3.18). Since the baseline vehicle was the pre-2000 model year F150, most of the steel used on the base was mild steel as was typical of this era of pickup truck.

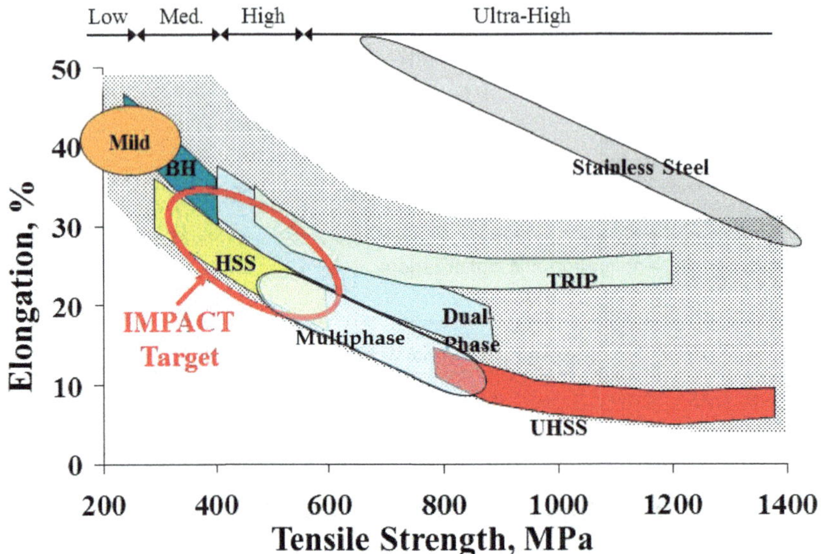

Figure 3.18 IMPACT Phase II material focus

For the IMPACT Phase II Cab and front end, 20% weight reduction was achieved (Figure 3.19). The figure shows the changes from the baseline in the top left to the final version in the bottom right.

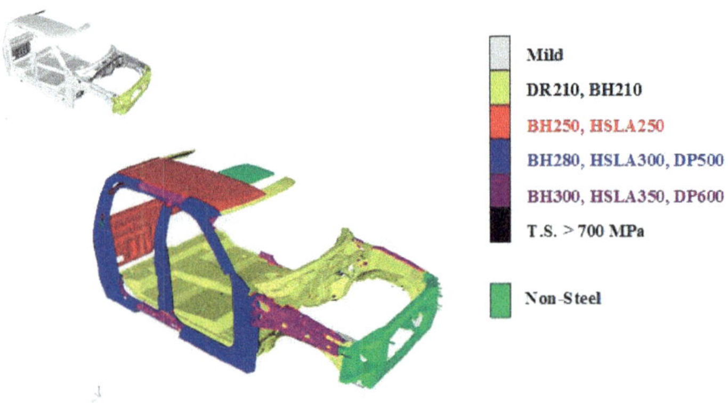

Figure 3.19 IMPACT Phase II cab and front end

The major weight reduction technologies used on the Phase II cab (Figure 3.20) were the following:

1. Structural adhesives were used extensively to recover the stiffness loss through metal thickness reduction.
2. DP 600 was used in the rocker and the three floor cross-members.
3. An LWB was used for the body side outer and C-pillar.
4. The radiator support was made of magnesium.

One of the limitations of the technologies that were considered for IMPACT Phase II was that the actual production vehicle that was being targeted was architecturally similar to the baseline vehicle. This meant that the amount of "stretch" that could be implemented for IMPACT Phase II was also limited. This was not true for IMPACT Phase III, which was targeted at the super duty truck. Two examples of a more radical approach to the Phase III cab were in the underbody (Figure 3.21) and the front end (Figure 3.22). The Phase III underbody benefitted with a complete architectural optimization, and DP 600 was used extensively. Two completely different architectures were considered for the Phase III front end (i.e., a stamped design and a hydroformed design). The stamped design achieved a 30% weight reduction from the baseline, but the hydroformed design was adopted for the production vehicle because it was the best combination of weight reduction and function. Also, DP 500 fenders were used in IMPACT Phase III.

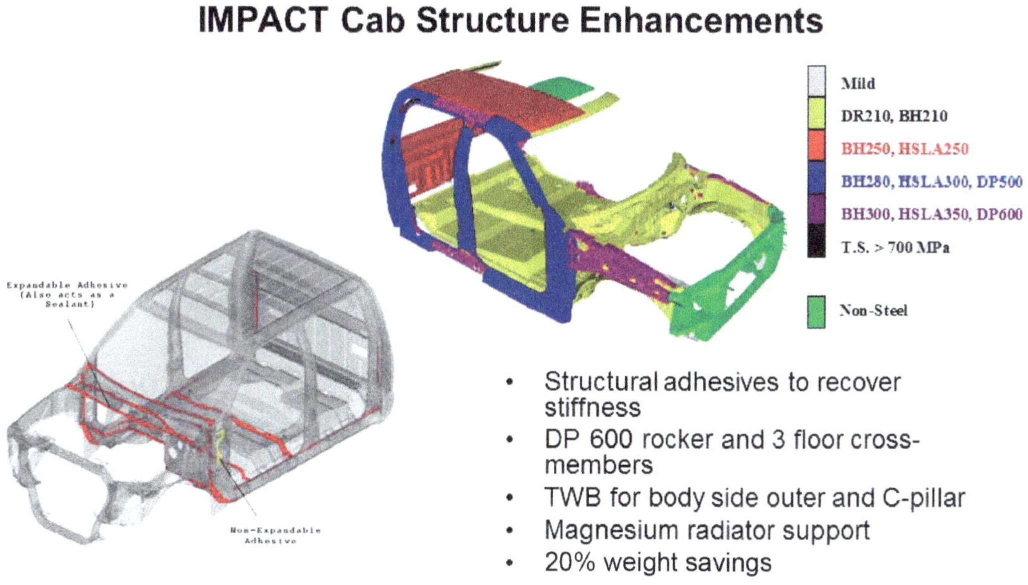

Figure 3.20 IMPACT Phase II cab technologies

Phase III Cab Underbody

- Phase II had limited architectural changes
- Phase III allowed complete architectural optimization
- DP 600 cross-members
- DP 600 longitudinal members
- Improved stiffness & durability
- 10% weight savings

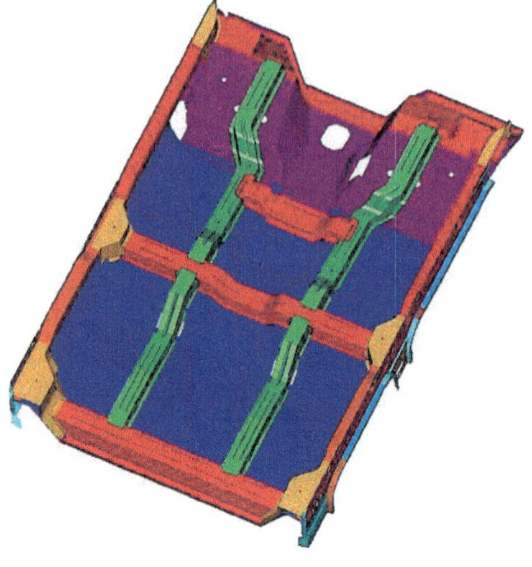

Figure 3.21 IMPACT Phase III underbody

Phase III Front End

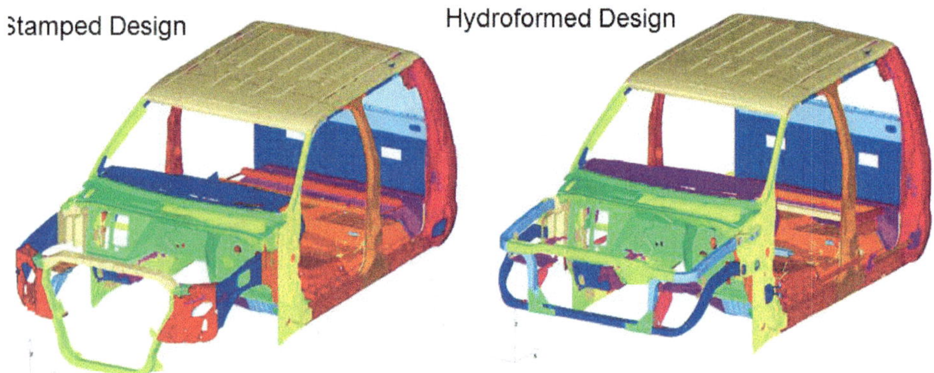

- Stamped DP 600 shotgun with design changes for stamping feasibility (30% weight reduction)
- Stamped DP 500 EG Fenders (20% weight reduction)
- Initiated hydroformed front end design for further design and package optimization

Figure 3.22 IMPACT Phase III front end

The Phase III pickup box was also radically redesigned (Figure 3.23). Essentially, the box inner was converted from a stamped design to a roll-formed design. This enabled the use of much higher strength steels at thinner gauges. The box outers were stamped out of thin gauge DP 500. The Phase III tailgate was redesigned with an extra strainer providing more stiffness, which in turn provided the opportunity for thickness reductions (Figure 3.24). Most of the tailgate was constructed out of DP steels. The tailgate was analyzed for formability and was judged to be easily stampable at the selected gauges (Figure 3.25).

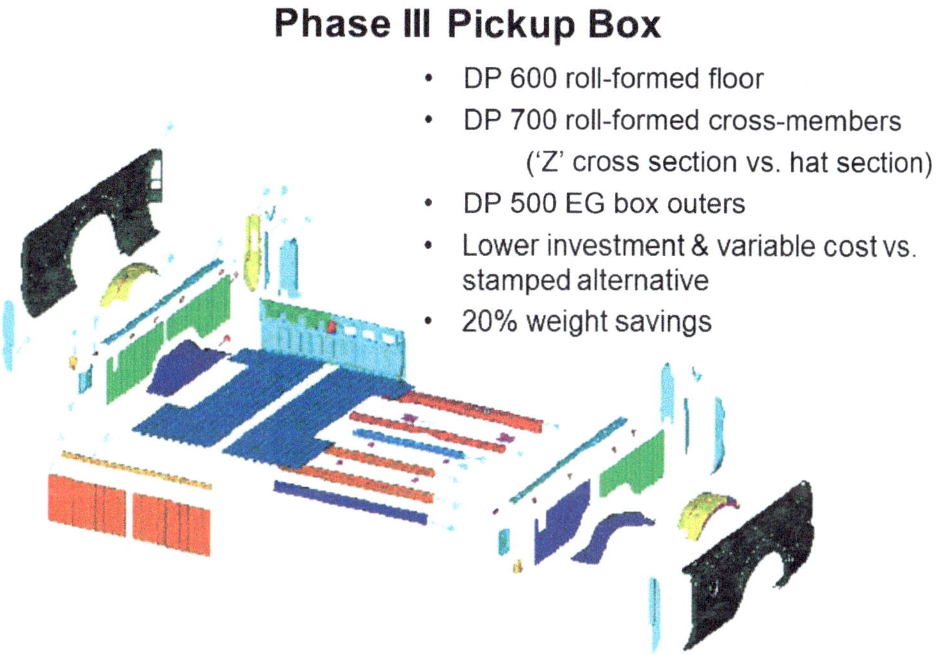

Phase III Pickup Box

- DP 600 roll-formed floor
- DP 700 roll-formed cross-members ('Z' cross section vs. hat section)
- DP 500 EG box outers
- Lower investment & variable cost vs. stamped alternative
- 20% weight savings

Figure 3.23 IMPACT Phase III pickup box

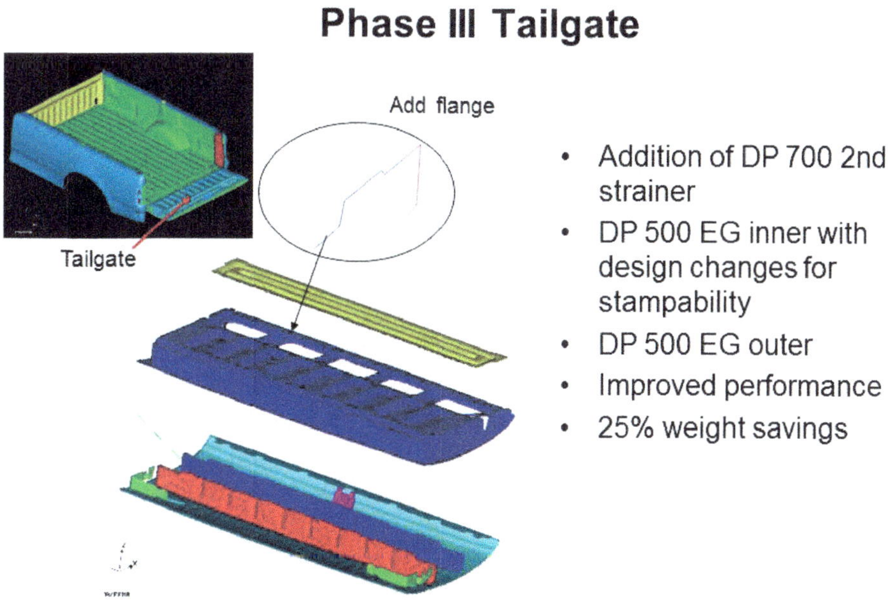

Phase III Tailgate

Add flange

Tailgate

- Addition of DP 700 2nd strainer
- DP 500 EG inner with design changes for stampability
- DP 500 EG outer
- Improved performance
- 25% weight savings

Figure 3.24 IMPACT Phase III tailgate

Phase III Tailgate

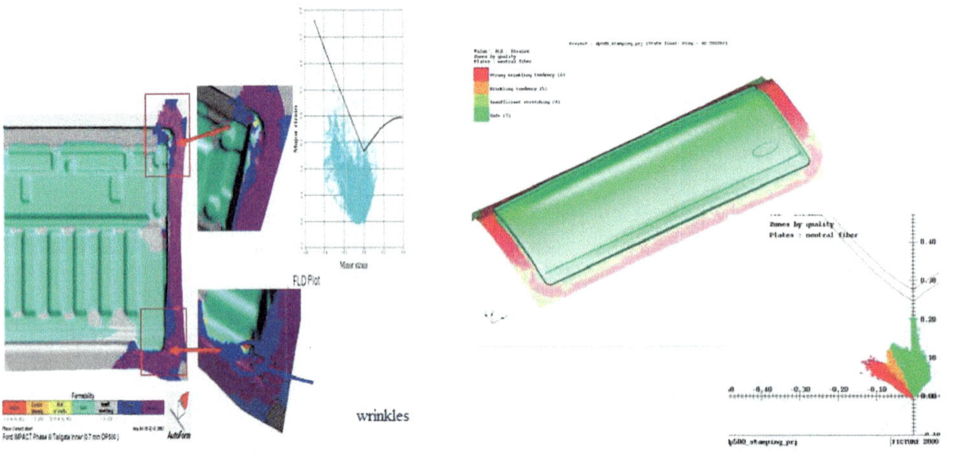

Formability Index Contour of 0.7 mm DP 500 Steel
For Tailgate Inner

Formability Index Contour of 0.7 mm DP 500 Steel
For Tailgate Outer

Figure 3.25 Tailgate formability analysis

Since only IMPACT Phase II had the 25% weight reduction requirement, as opposed to Phase III, more lightweight alternative materials were used in Phase II. This is what drove the magnesium radiator support in Phase II, and it also drove the aluminum cross-members, knuckles, and lower control arms on the Phase II frame (Figure 3.26). The Phase III frame was made entirely out of steel.

1. The rear side rails were converted from 6.7 mm to 5.5 mm hot-rolled DP 600 for a 17.9% weight savings.
2. DP 600 was not used for the front rails because of crush considerations.
3. The Phase III frame was built and installed in a vehicle at 13,000 lb gross vehicle weight (GVW). The vehicle was tested at the manufacturer's durability course and passed all requirements.

One of the areas of focus on the Phase III frame was the design of the transmission cross-member. The baseline design was a heavy, complicated design with nine major pieces (Figure 3.27). Two different lightweight designs were developed. One was a hydroformed design, which reduced the weight by 28%, and another was a sheet metal stamped design, which reduced the weight by 23% (Figure 3.28). The hydroformed version was designed with varying thicknesses throughout the length of the cross-member, where the thicker sections were in the areas in which more strength was needed. The stamped version reduced the number of major parts from nine to five while increasing the cross-sectional area to provide more overall stiffness, using thinner gauge parts.

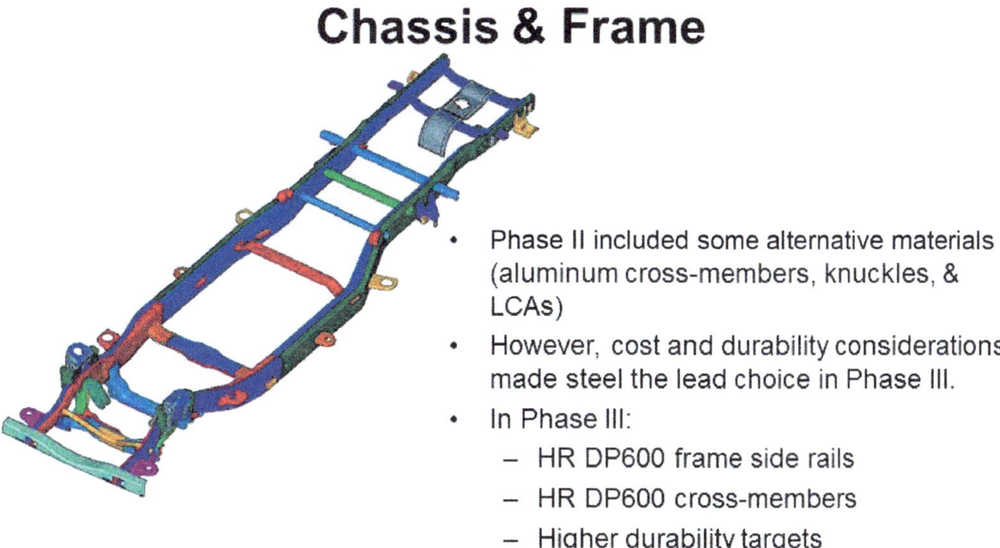

Figure 3.26 IMPACT chassis and frame

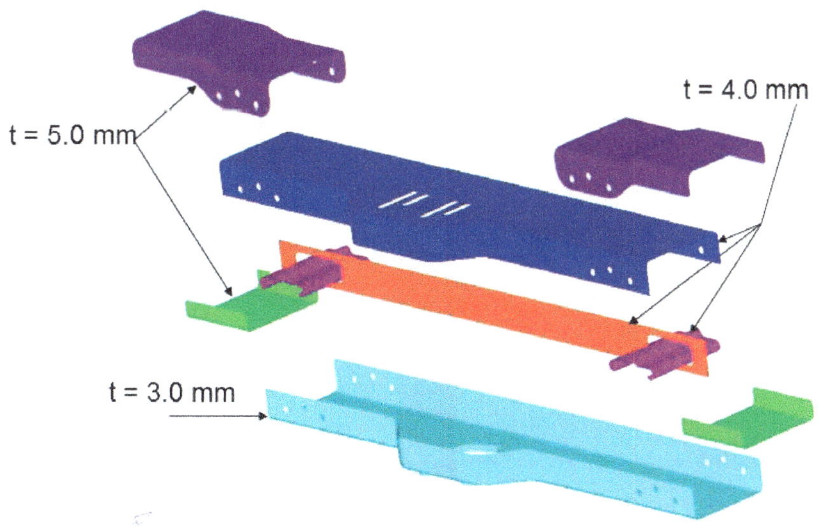

Figure 3.27 Baseline transmission cross-member

HSS Transmission Cross-Members—Hydroformed & Clam Shell Design

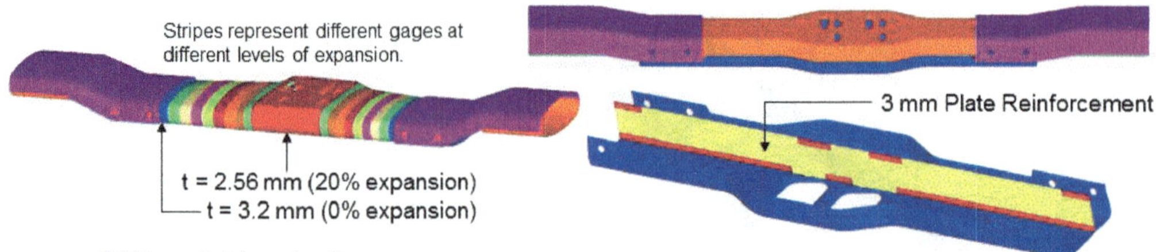

- 28% weight reduction
- No overlapping flanges reduces weight.
- 70% less MIG welding required.
- Eliminates one set of dies (1 part).
- Reinforcement for more stiffness.

- 23% weight reduction
- Increased cross section.
- Partially overlapping flanges to reduce weight.

			Bracket	Center	Static Stiffness (N/mm)		
	Mass (kg)	Freq. (Hz)	Gages (mm)	Gage(s) (mm)	x	y	z
Baseline	27	217	5	3.5	2767	5405	3665
Hydroformed	19.5	250	3.5	3.2	2653	4037	2896
New Clam	19.7	250	3.6	3	2747	4509	2343
New Clam	20.9	250	3.6	3	5241	7241	3386
with 3 mm Reinf.				Percent Increase	48%	38%	31%

Figure 3.28 Weight reduced transmission cross-members

IMPACT Phase II was driven to deliver the high value (i.e., weight reduction divided by variable cost) solution, which would also be consistent with achieving a 25% weight reduction. Typically, the steel alternative when sufficient for achieving the 25% goal was also the high-value solution. The percentage of steel remained the same in the final design when compared with the base, while the strength of the steel moved up considerably in the final design (Figure 3.29). The variable cost of the steel used actually decreased when going to the

IMPACT Material Utilization by Weight

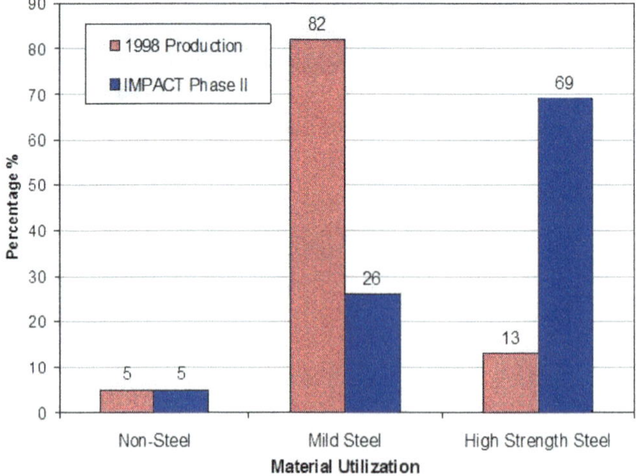

Includes Body Structures, Frame & Wheels, and Closures

Figure 3.29 Material utilization by weight

final design (Figure 3.30). This was accomplished by making very conservative "step-ups" in grade. By this strategy, the cost reductions provided by the thickness reductions would more than offset the increased cost of the new grades. This strategy also ensured that the new tooling costs would be minimized. Also, to reduce costs, steel designs were continuously refined from a weight perspective so that in the end the steel alternative could displace an alternative material for a given subsystem (Figure 3.31).

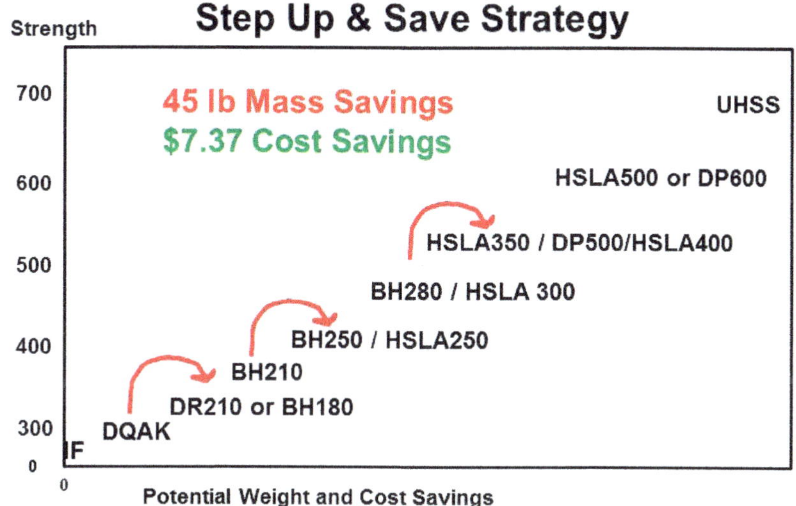

LESSON:
Strength step-up and gauge reduction strategy may be used with existing tooling.

Figure 3.30 Step up and save strategy

IMPACT PHASE II
Alternative Materials to Steel Conversions

Application	Start Design	End Design
Engine Mounts	Metal Matrix Composite	HSS Brackets
Trans. Mounts	Metal Matrix Composite	HSS
Muffler	Titanium	Stainless
Tow Hooks	Aluminum	HSS (Forging)
Skid Plates	Composite	HSS
Wheels	Aluminum	Kuhl Steel Wheel
Seats	Aluminum / Magnesium	HSS
Fender Outers	Aluminum	HSS
Front Door Assy	Aluminum	Spaceframe Steel Design
PUB BSO	SMC	HSS
Front Bumper	Aluminum	Stainless / HSS
Rear Bumper	Aluminum	Stainless / HSS

Figure 3.31 Alternative materials to steel conversions

In summary, the IMPACT Project delivered the following:

1. Extensive CAE modeling to drive weight efficient designs
2. A road map to achieve 25% weight reduction at carryover attributes
3. Fully functional vehicles that demonstrated compatible attribute performance
4. An application roadmap for steel grade usages throughout the body and chassis
5. A weight technology cascade plan to drive technology consideration across other vehicles
6. Manufacturing awareness of the future materials direction
7. Over 60% of the technologies piloted on the IMPACT project were implemented on the F150, Super Duty, or other Ford platform

3.3 Body Structure Safety Application

The next application that will be discussed is the Auto/Steel Partnership's Mass Efficient Architecture for Roof Strength (MEARS) Project, which was started at the beginning of 2006 in anticipation of the new FMVSS 216 roof strength regulation. Ford donated their current F150 Super Duty design properties for the study. The Super Duty was chosen because unlike the regular cab and the crew cab there was no B-pillar (i.e., the pillar between the front and rear side glass), which meant that the super cab presented the most difficult condition for passing the new roof crush. Also, the National Highway Traffic & Safety Administration let it be known that the new FMVSS 216 would be much harder on pickups than the previous law. The author led the project at the A/SP until he retired from Ford at the end of 2006, and continued to attend the meetings in 2007 as a representative of the vendor who was chosen to perform the initial study. After the author retired, Shawn Morgans from Ford took over the leadership of the project. The description of the project, which will be used here, comes in part from Shawn's presentation of the project at the 2009 Great Designs in Steel (GDIS) Conference. This project demonstrates how AHSS technology in conjunction with other enabling technologies can be executed to minimize the weight impact of a full vehicle crash event. It also provides an understanding of the cost trade-offs of different approaches for meeting a specific weight target and provides an understanding of how large a role vehicle architecture makes in achieving a weight target. The project plan was to:

1. Develop a test-correlated finite-element model to evaluate several different concepts and assess the sensitivity of different structural parts.
2. Develop multiple concepts, including stamping intensive, hydroform intensive, and structural insert intensive designs.
3. Optimize the concepts to reduce mass.
4. Compare the concepts from a mass, cost, and manufacturability point of view and choose a final design.
5. Fine-tune and optimize the final design.

Figure 3.32 shows a schematic of the process in which the first step was to validate and improve upon the CAE model, which was supplied by Ford. The next step was to create a reduced turnaround model, which would be amenable to extensive optimization. The three major concept categories included a hydroform intensive design, a stamping intensive design, and a stamped design with structural inserts. The concepts would then be compared and the concept that showed the most promise would be further optimized.

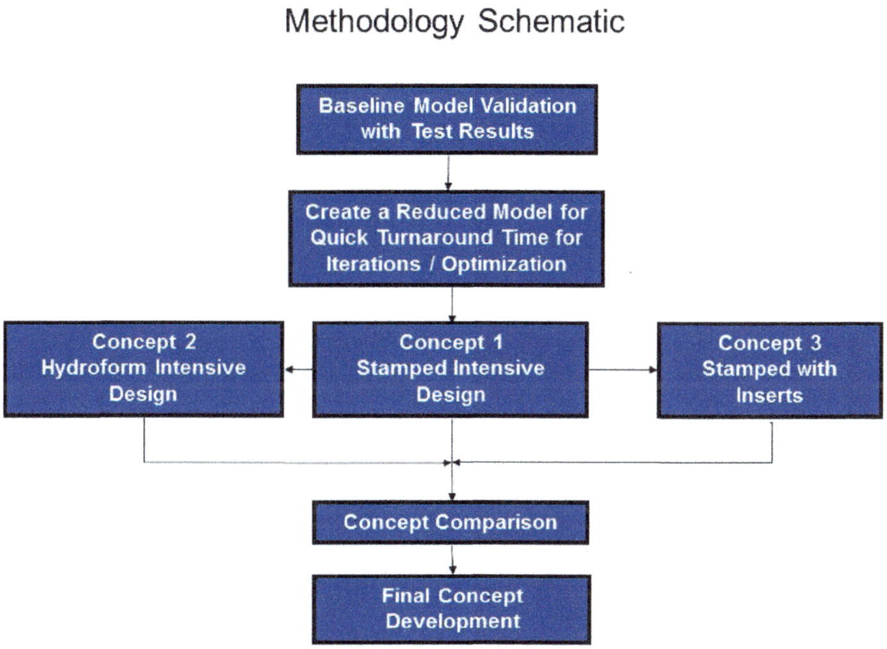

Methodology Schematic

Baseline Model Validation with Test Results

Create a Reduced Model for Quick Turnaround Time for Iterations / Optimization

Concept 2 Hydroform Intensive Design

Concept 1 Stamped Intensive Design

Concept 3 Stamped with Inserts

Concept Comparison

Final Concept Development

Figure 3.32 Methodology schematic

The goal of the project was to achieve a load of 20% above what was thought to be NHTSA's new regulation at the time that the project started. At the time, it was thought that by 2011 the new FMVSS 216 would require that a platen, which would contact the roof diagonally, would reach a load of 2.5 Gs before impacting a specific head form for all cars and light trucks. To be consistent with this regulation and to provide our safety factor, our goal was to achieve 3 Gs before 4.5 in. displacement was reached. NHTSA actually implemented different requirements than what was thought at the time, but our intent here is not to go into a discussion of the specific regulation but to explore technologies that could address a more stringent requirement than what was then in place.

In terms of the baseline CAE model correlation and sensitivity, the results are shown in Figure 3.33. The correlation with test was judged to be very good, and the critical sections were shown to be at the base of the A-pillar (i.e., the A-pillars are the vertical pillars on either side of the windshield), the A-pillar to roof joint, and the C-pillar (i.e., the rear-most vertical pillar).

Baseline Model & Test Correlation

A B-pillar less pick-up truck was chosen for this study.
The finite-element model was fine-tuned to correlate the results with the rest data of the old regulation.

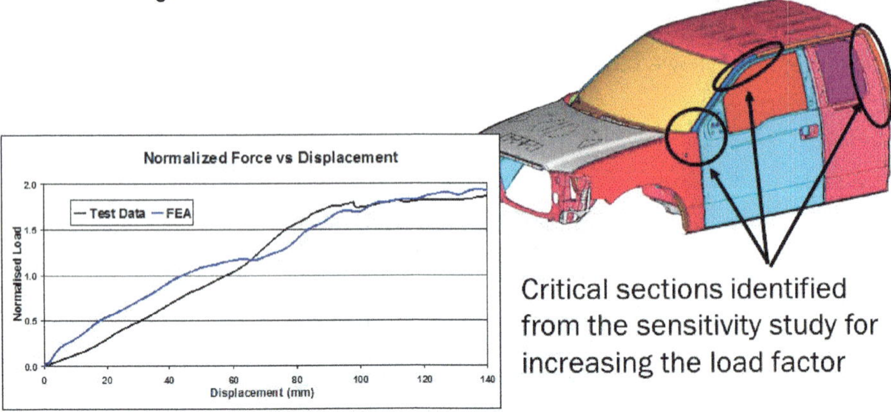

Figure 3.33 Baseline model and test correlation

The modes of buckling for the critical sections are shown in Figure 3.34. The locations of buckling for the top and bottom of the A-pillar are shown on the left side of the figure, and the C-pillar buckling movement is shown in the top right. The hinge pillar was also buckling, just below the A-pillar attachment.

Baseline Design – Critical Sections

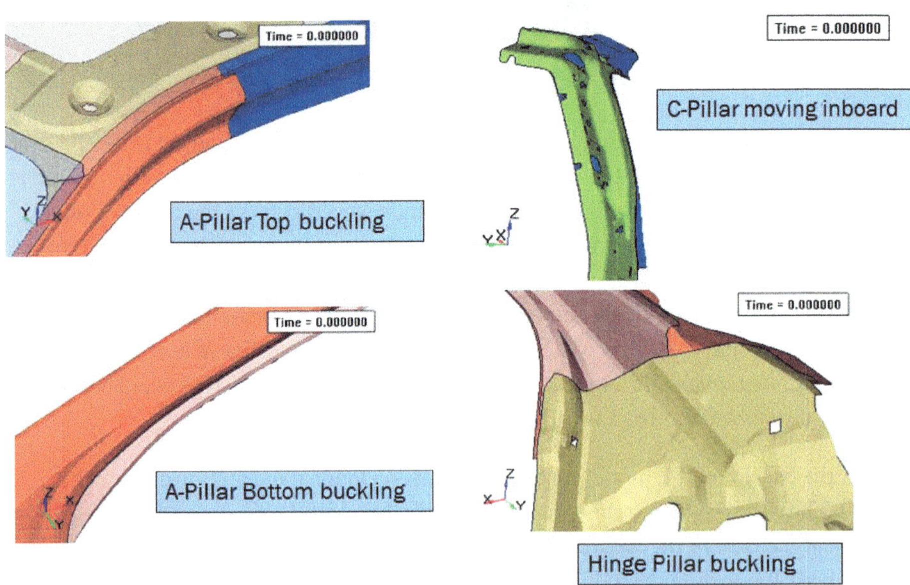

Figure 3.34 Baseline design—Critical sections

To create the stamping intensive design, the team focused on the critical sections identified in the previous step of the work plan (Figure 3.35). The A-pillar geometry was modified to give better support throughout the full length of the A-pillar. The top of the hinge pillar inner was modified to give a better connection with the bottom of the A-pillar. A C-pillar reinforcement was added to improve the bending stiffness of the C-pillar. Also, a rear header outer was added to give more rigidity to the left side of the greenhouse by providing a better load path to the right-hand side of the greenhouse. AHSS were used for the A-pillar, roof rail, and C-pillar parts.

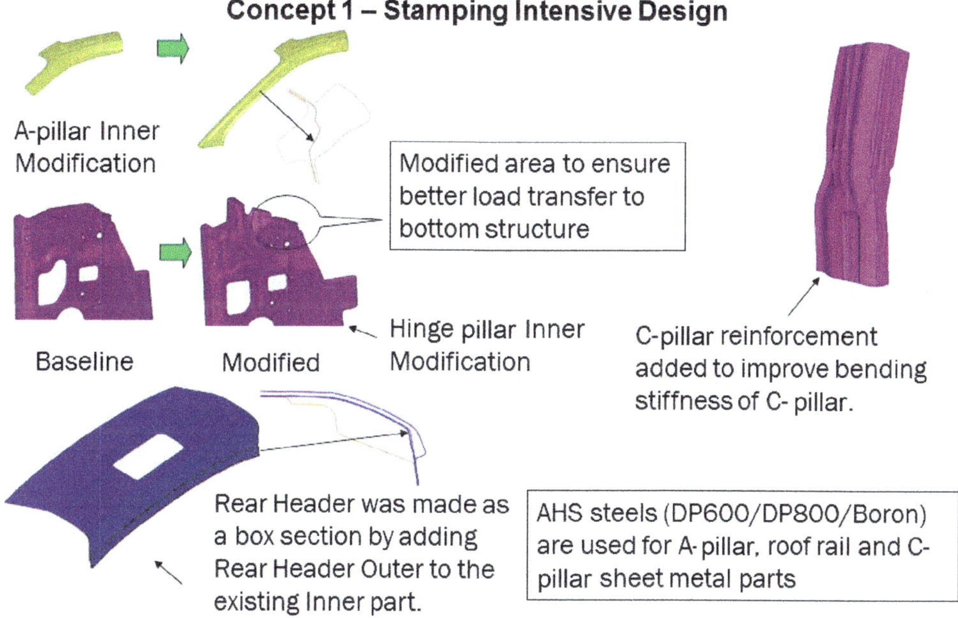

Figure 3.35 Concept 1—Stamping intensive design

Figure 3.36 shows the hydroform intensive design. This design actually has a total of three hydroformed tubes: Each A-pillar has a hydroformed tube that runs from the top of the hinge pillar, through the A-pillar and roof rail to the C-pillar. There is also a hydroformed tube that starts at the bottom of the C-pillar and runs across the rear header and down the C-pillar on the other side. In this design, the A-pillar reinforcements and roof rail outer were removed from the previous concept. The tubes were DP 800.

For the structural insert intensive concept, we actually came up with four designs (Figure 3.37). For the first two designs, two different versions were proposed, using steel inserts that were added to critical places in the greenhouse. The third design used polyurethane foam inserts, which were to be injected into critical cavities of the greenhouse. In the fourth design, nylon inserts would be dropped into the critical areas of the greenhouse. The nylon inserts utilize heat-activated expanding adhesives in a nylon carrier cage type structure.

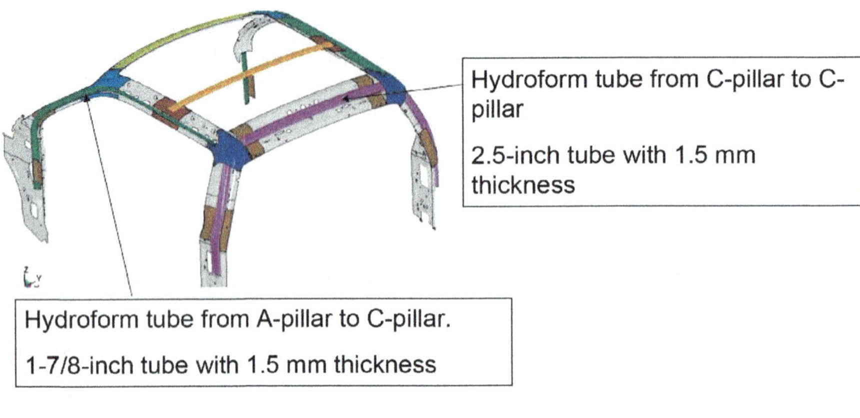

Figure 3.36 Concept 2—Hydroform intensive design

All the designs were subjected to extensive computer optimization, and a comparison of the various designs is shown in Figure 3.38. A vertical line is drawn at 4.5 in. (114.3 mm). Again the baseline test data and CAE model results are shown as reaching about 1.8 Gs at the 4.5 in. line. All the other designs reach at least 3Gs at that displacement, with the beta foam design reaching the highest load.

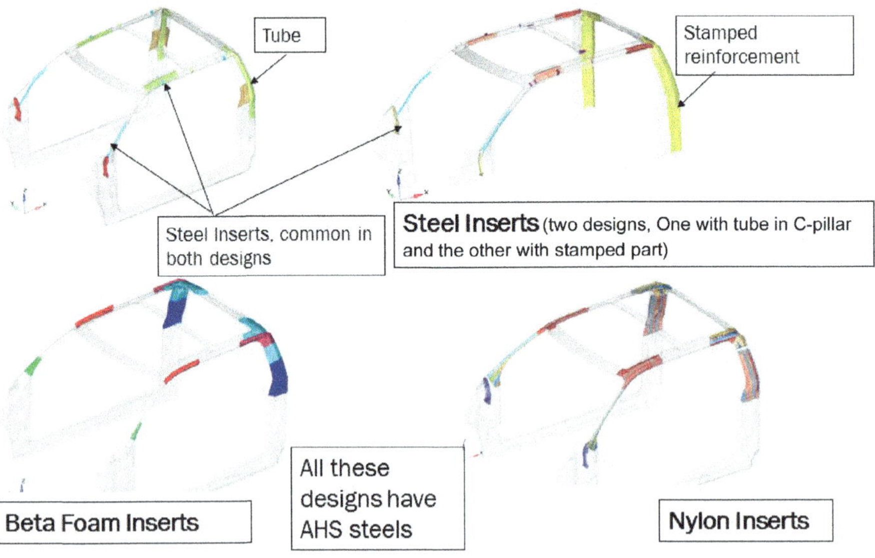

Figure 3.37 Structural insert strategies

Comparison of Concepts

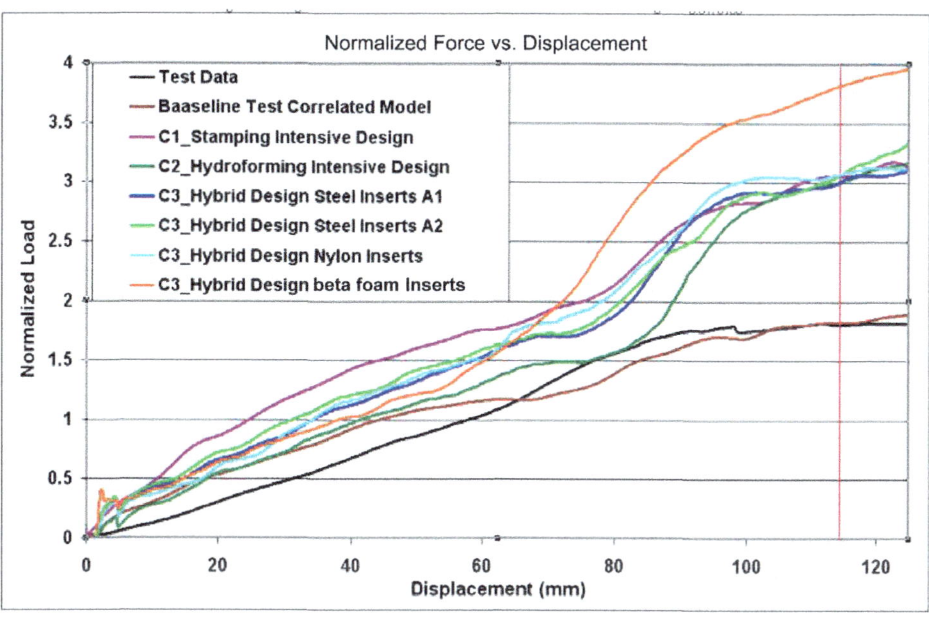

Force displacement curves for all the conpects

Figure 3.38 Results of the various concepts

The results in tabular form are shown in Figure 3.39. Note that the result for the beta foam concept, which was shown in the previous figure, was under the assumption that the foam was fully bonded to the sheet metal. Under the assumption that there was no bonding, the beta foam only achieved 2.95 Gs at 4.5 in. Note also that all the concepts used AHSS. The table lists the load, mass added, and cost. Then the designs were assigned a rating for mass, cost, manufacturability, and repair. Finally, based on the rating, each design was assigned a weighted rating, based on the rating for mass, cost, manufacturability, and repair. Design 3A2 and 3B tied, but the team selected the nylon insert design for further optimization. This design met the load factor requirement with increases in mass of 7.5 kg and cost of $80.

A description of how the selected design was further optimized is shown in Figure 3.40. The optimization was done in two phases. In the first phase, shape, section, and size optimizations, learned from the other concepts, were tried. For instance, the rear header was modified as shown in the figure. The nylon inserts were also manually optimized in this phase. The mass increase was reduced to 4.5 kg in this phase. In phase 2, further optimization was done using Heeds optimization.

Comparison of Concepts Cont'd

S.No	Concept	Load Factor	Mass [kgs]	Cost	Rating				Weighted Rating
					Mass	Cost	Manufac-turability	Repair	
Weight Factor --------------------------->					4	3	2	1	
1	Stamping Intensive	3.06	17.6	$108	1	1	4	3	18
2	Hydroform intensive	3.00	10.5	$79	4	3	3	3	34
3 A1	Steel Inserts-Tube in C-pillar	3.00	14.9	$79	2	3	3	4	27
3 A2	Steel Inserts-Stamped C-pillar Rnf	3.06	13.8	$67	3	5	4	4	39
3. B	Nylon Inserts (Drop-in)	3.06	7.5	$80	5	3	4	2	39
3. C	Beta foam (Injected)	2.95-3.82*	8.4	$78	5	3	2	2	35

**This lower and upper bounds of the load factors refers to no-bonding and 100% bonding between sheet metal and Beta Foam*

1. A total of six concepts were generated and compared.
2. AHSS steel is used in all the concepts.
3. Hybrid design with AHSS and nylon inserts is chosen as the final design. This design has the least mass increase of 7.5 kg over the baseline model.
4. Further optimization was carried out on this hybrid design.

Figure 3.39 Comparison of concepts

Optimization of Selected design

- Optimization on the selected Design was done in two phases
- Shape, section, and size optimizations learned from the other concepts are tried in phase 1. Nylon inserts are also manually optimized in this phase.
- The mass increase after the phase 1 optimization is 4.5 kg.
- Further automated optimization was done in phase-2 using Heeds optimization

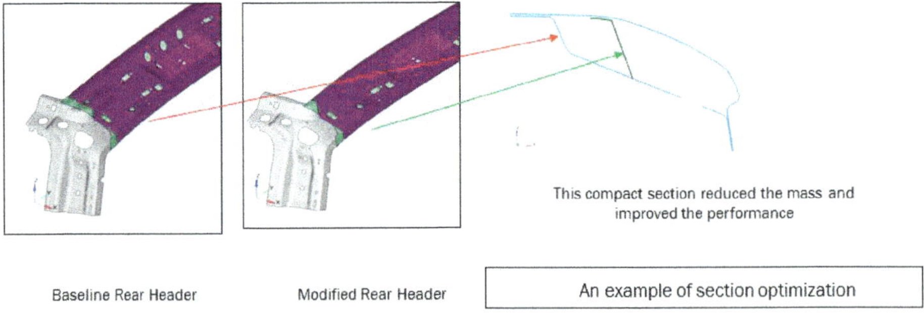

Figure 3.40 Further optimization of the nylon insert design

As shown in Figure 3.41, the final optimized design came out as only 1.2 kg heavier than the baseline, with a significant increase in performance.

Final Design Performance

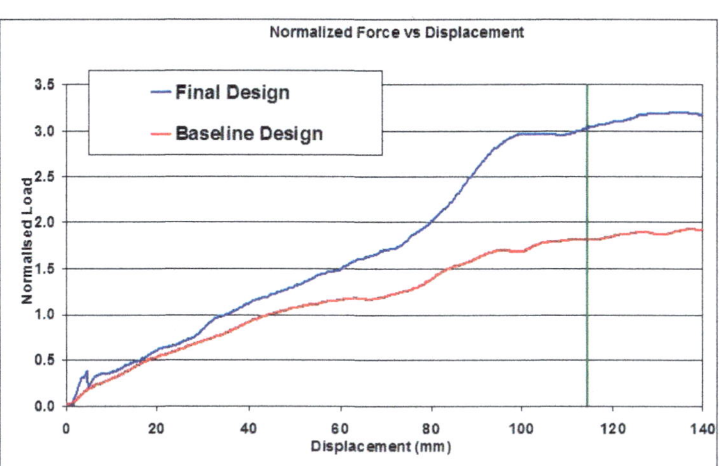

Optimized hybrid Design weighs only 1.2 kg more as compared to the baseline design.

Performance improved by 66%

Figure 3.41 Final nylon insert optimization

The next two figures shows the details of the steel modifications, which were made to the final nylon insert concept. Figure 3.42 shows the body side modifications, which included a LWD body side outer and C-pillar inner and hot-stamped A-pillar reinforcement and roof rail reinforcement. Figure 3.43 shows the lateral roof elements and the rear door design. Figure 3.44 shows the final configuration of the actual nylon inserts.

At this point in the project, the following were delivered:

1. Mass efficient design solutions would allow compliance to the proposed new roof strength regulation; FMVSS 216 for the most difficult vehicle configuration (i.e., B-pillar-less pickup truck) were developed.
2. Several concepts, using AHSS materials and architectural elements like LWB, hydro-forming, hot stamping, and structural foam were used in the developed concepts.
3. The hybrid design consisting of AHSS materials, steel, and nylon inserts was chosen for final optimization.
4. An optimized vehicle architecture with only 1.2 kg mass increase was developed, which met the target deflection of less than 4.5 in. under three times the vehicle weight.
5. This results in an increase of more than 70% in the performance as compared with the baseline design.

Final Design Details – Sheet Metal Parts

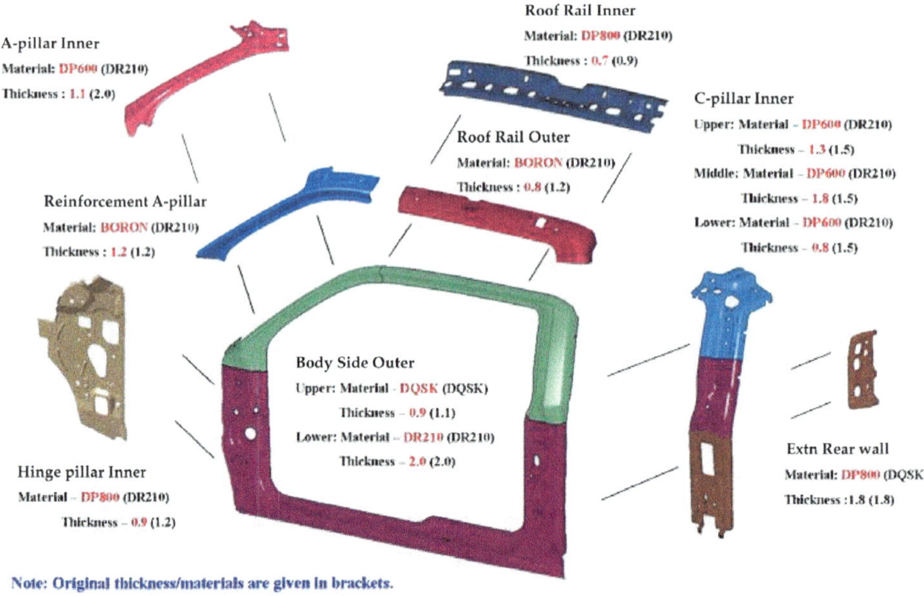

Figure 3.42 Body side changes for the nylon insert design

Final Design Details – Sheet Metal Parts Cont'd

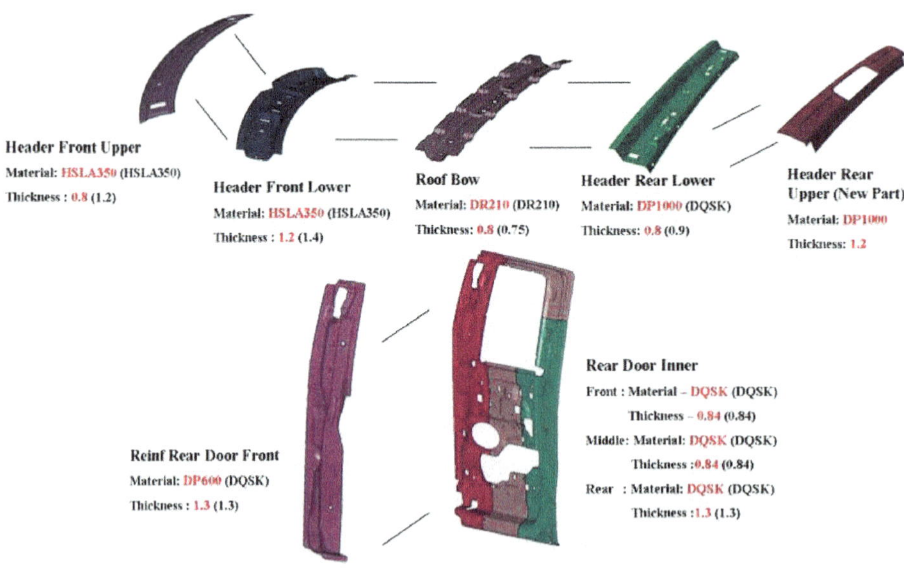

Figure 3.43 Lateral roof elements and rear door changes for the final nylon insert design

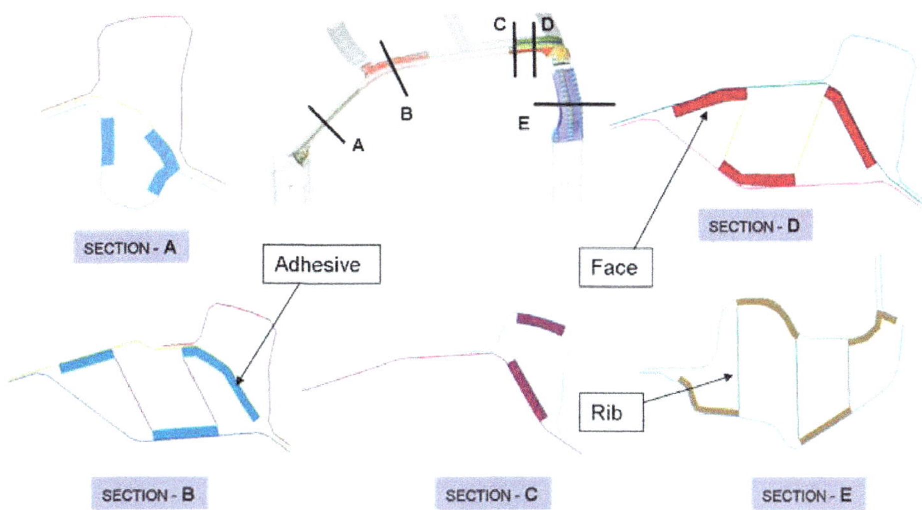

Figure 3.44 Final configuration of nylon inserts

This was the status of the project at the end of 2007. At that point, the MEARS team decided to continue the study into 2008 and to start a major Phase II study with the following tasks:

1. Continue the optimization of the final Phase I model.
2. Demonstrate the correlation between CAE and the actual test of composite reinforcements.
3. Investigate the impact of component geometry on roof strength performance.
4. Utilize the continuous joining methods to enhance performance.
5. Conduct feasibility studies.
6. Develop costs for proposed changes/adds.

The A/SP MEARS team worked on Phase II for most of 2008 and did achieve an improvement in the metrics for the Phase I project. Figure 3.45 shows the final design. The following were the conclusions of the project after Phase II:

1. Phase I of the study concluded that multiple solutions achieved the targets established by the initial FMVSS 216 initial targets.
2. The Phase I team determined that the use of composite inserts in conjunction with AHSS and ultra-high-strength steels provided the best design option.
3. Phase II of the project continued the optimization of the final design from Phase I and determined the following:
   a. A load of 3 Gs was achievable with a hybrid structure (steel and composite reinforcements).

b. Optimization of the spot weld count can provide benefits to roof strength performance.

c. A 4.4 kg mass savings over the baseline donor vehicle body structure was possible while increasing the load carrying capacity by 60%.

d. The estimated cost increase of the Phase II design was $70 per vehicle.

4. Correlation between CAE and physical testing for composite reinforcements was established.

5. Manufacturing feasibility of the design was confirmed.

6. The use of linear optimization early in the development of a program provides insight into the required structural changes.

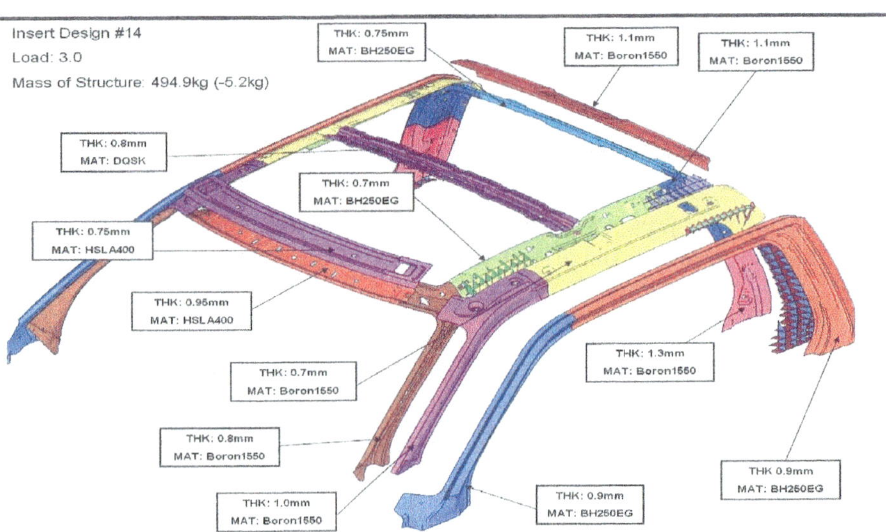

Figure 3.45 MEARS Phase II optimized design

All seemed to be going well for the MEARS project, until the 2009 GDIS. At that conference, Ford brought a physical property for the next generation F150. This vehicle had a hydroforms section on each side of the vehicle, which extended from the hinge pillar, through the A-pillar, the roof rail, and the C-pillar. The extension of the hydroform through the C-pillar differed from the MEARS hydroform design in that the MEARS hydroforms stopped at the top of the C-pillar and had a separate hydroform going laterally from C-pillar to C-pillar. Shawn Morgans also gave a separate presentation at the GDIS explaining the Ford design, which was claimed to be superior in metrics to the MEARS design. The confusion was explained by Shawn Morgans during the question and answer period following the Ford F150 presentation. In the first place, Ford favored their hydroform configuration versus the configuration chosen by the MEARS team. Also, one of the critical assumptions at the beginning of the MEARS project was that if a hydroform design were used, the body side assembly would still need a body side inner to attach the interior body panels to. This was actually the industry standard assumption at the time that the MEARS project was begun. The Ford hydroform design did not have a body side inner, as the interior panels were attached directly to the hydroform beam.

3.4 Auto/Steel Partnership Closure Application

This section will review the history of the Auto/Steel Partnership Lightweight Closure Project from inception in 2000 through completion in late 2005. After reading this section, you should be able to

1. Describe the most efficient architecture for doors
2. Specify the types of AHSS that would be appropriate for doors
3. Identify how to overcome forming impediments for door panels

The project was formed to develop and prove out door lightweighting concepts. The assumption was that the studies concepts would be brought closer to production feasibility when compared with other non-A/SP activities (e.g., American Iron and Steel Institute Ultralight Steel Auto Body Consortium). The first two phases of this project were primarily concerned with modifications to the inner door structure, and the third and final stage was concerned with material modifications to the door outer.

The team (Figure 3.46) that worked on this project consisted of members from the participating companies of the A/SP as well as members of Oxford Automotive, which was

Figure 3.46 A/SP lightweight closure team

the major contractor for the execution of our workplan. Several other companies (Altair Engineering, J.F. Hubert Enterprises, Henkel Technologies, Lord Corporation, and Icon Creative Technology Group) also contributed to this study. I also led this project. Fifty percent of the funding for the third and final stage of this project was provided by the U.S. Department of Energy through the Freedom Car and USCAR programs.

The following were the overall constraints for the lightweight closure team:

1. Find mass reduction possibilities.
2. Obtain a contributed door design, which would be benchmarked and optimized using innovative approaches.
3. Maintain the vehicle constraints and attributes that the door was originally designed for while maintaining the original vehicle level architecture.
4. Also, the innovations would have to be commercially feasible.

The objective for the first phase of the project, which lasted from 2000 through 2001, was to achieve a 25% weight reduction in the benchmark door-in-white at less than a $0.70 variable cost increase per pound saved, by investigating several door architectures for light-weighting, including "patch" welding, LWB, and hydroformed components. This led to the development and refining of several door concepts. In formulating this project, the project team and Oxford agreed that we would evaluate the benchmark/baseline door, and four concepts based on a traditional construction, an optimized LWB construction, hydroform alternatives, and a multipiece door-inner construction (Figure 3.47). The major focus of this study would be on the door-inner construction, as that is the heaviest piece of the door. The upper picture in Figure 3.47 shows the benchmark door. As you can see, the benchmark door inner was an LWB. Also in this figure, you can see the other major structural pieces of the door (i.e., the door outer, the crash beam, the beltline reinforcement, the hinge and latch reinforcements, and the glass guides).

The two concepts that performed the best under our selection criteria are shown in Figure 3.48, with the multipiece being the clear winner, followed by the multipiece with a hydroform armature included. The concept of the multipiece was that the door inner would initially be stamped as six separate pieces and then the pieces would be joined after stamping versus before stamping as in the LWB. This meant that the thickness could be optimized over six regions of the door inner. In the hybrid multipiece/hydroform shown in the figure, the hydroform becomes the major structural member with further thickness reductions possible along with the elimination of a separate crash beam.

In terms of quantitative results, the multipiece met our variable cost objective, and it provided a 16.5% reduction in weight. Though it was significantly lower than the 25% weight objective, it still gave more weight reduction than the other alternatives. However, the A/SP Joint Policy Board rejected this alternative because it represented too much dimensional risk for a mass-produced door. At the time of this study, door NVH (e.g., squeaks and rattles, wind noise, and so on) and opening/closing efforts were high on the list of concerns for all the auto manufacturers, and these were directly related to dimensionality.

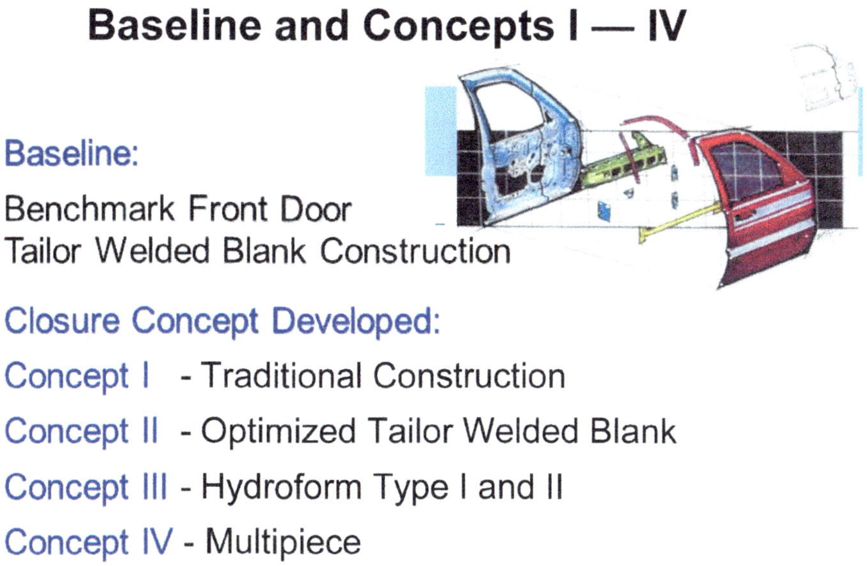

Baseline and Concepts I — IV

Baseline:

Benchmark Front Door
Tailor Welded Blank Construction

Closure Concept Developed:

Concept I - Traditional Construction

Concept II - Optimized Tailor Welded Blank

Concept III - Hydroform Type I and II

Concept IV - Multipiece

Figure 3.47 Baseline and concepts I–IV for Phase I

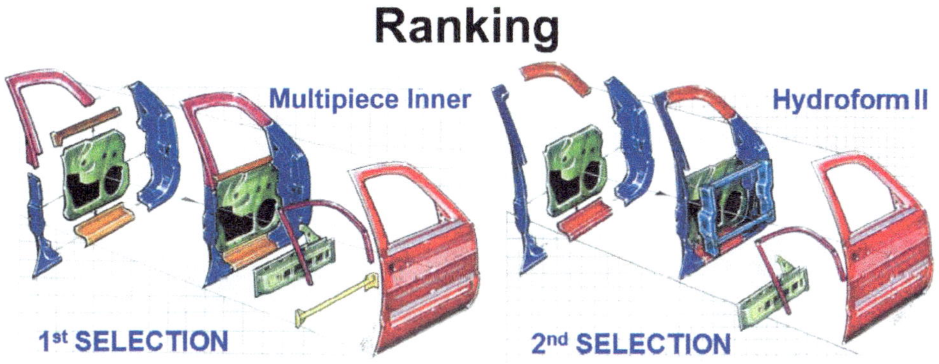

Ranking

Multipiece Inner

Hydroform II

1st SELECTION

2nd SELECTION

Figure 3.48 Ranking of the two leading Phase I designs

The directions that were given to the team by the Joint Policy Board in 2001 was to focus on what could be achieved in a door-inner design by developing several different alternatives, focused around LWB and hydroform concepts. Phase II was to extend through the end of 2002. Based on this direction, the objective for Phase II was to focus all the design efforts on the door inner, because as the heaviest portion, it presented the greatest opportunity. It also provided the greatest challenges because of various technical barriers. For instance the door-inner design is primarily stiffness driven, which gave no advantage to the use of AHSS. No specific weight reduction target was set for this phase, but a 15% decrease was thought to be realistic. The technical approach was to develop multiple door-inner designs using several variants of the LWB and hydroform alternatives explored in Phase I.

In terms of the LWB designs, which were tried, an illustration is shown in Figure 3.49. The baseline door is shown in the center of the figure, and the two LWB concepts, which were optimized, are shown to the left of the figure. Instead of using two sheets of steel, which are welded along a single linear weld line, the upper alternative used four sheets of steel, welded along three weld lines. The alternative on the bottom left used two sheets of steel, welded with a nonlinear weld line. One of the reasons that these alternatives were chosen was to minimize the forward part of the LWB in the baseline, because this part was three times thicker than the rearward part. This is because door sag is typically the hardest of the attributes to meet and requires a very stiff forward part of the door. The other two alternatives, which were optimized, were "patch blank" concepts. In a patch blank, the door inner is a single piece, with pieces of steel welded on the door inner blank, where you would need more stiffness or strength, before stamping. Of course, you would only be able to weld the pieces at a few points so that the pieces could slide with respect to the main blank during stamping.

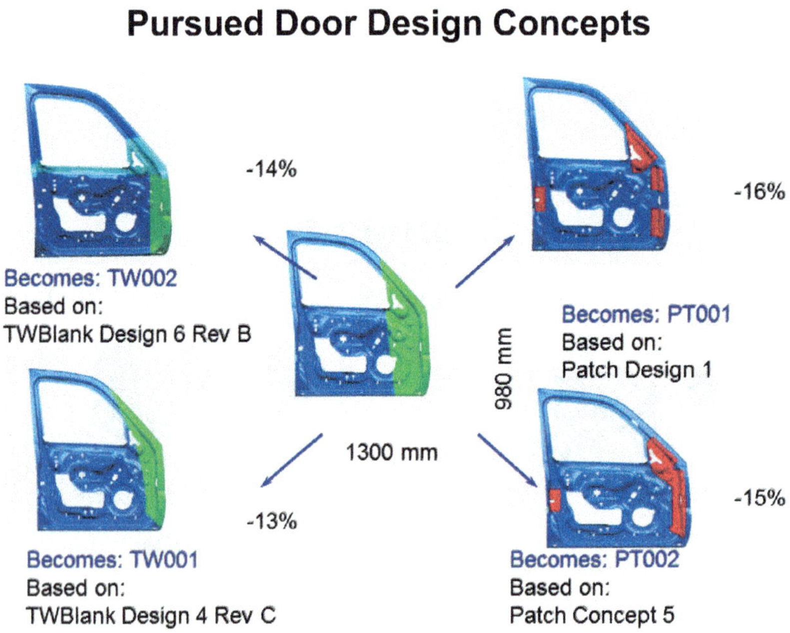

Pursued Door Design Concepts

-14%
-16%
Becomes: TW002
Based on:
TWBlank Design 6 Rev B
Becomes: PT001
Based on:
Patch Design 1
980 mm
1300 mm
-15%
-13%
Becomes: TW001
Based on:
TWBlank Design 4 Rev C
Becomes: PT002
Based on:
Patch Concept 5

Figure 3.49 Laser-welded and patch blank concepts

The two new hydroform concepts, which were tried in Phase II, are shown in Figure 3.50. On the left is a small hydroform, shown in blue at the rear of the door, with longitudinal tubes for the beltline reinforcement and the side impact beam. The second concept has a hydroform beam at the front and rear of the door, tying together the beltline reinforcement and the side impact beam. Since the major problems with the previously explored hydroform concepts had been too much weight and too high a cost, an attempt was made to simplify the hydroformed elements, as shown in this illustration, to get the "most bang for the buck."

However, at the end of Phase II, none of the optimized designs yielded very impressive results. Looking at the chart in Figure 3.51, which includes the multipiece, none of the Phase

II results were as good as the Phase I multipiece. Note that the patch blank concepts did not meet the sag performance target because of the inability for the patch blank to carry a shear load from the patches to the main blank.

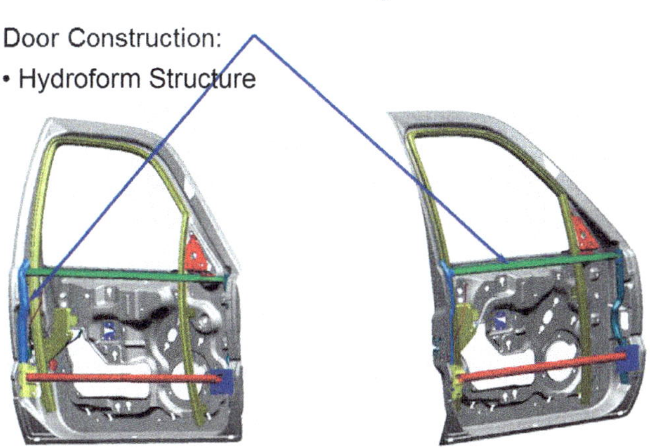

Figure 3.50 Phase II hydroform concepts

Performance Data

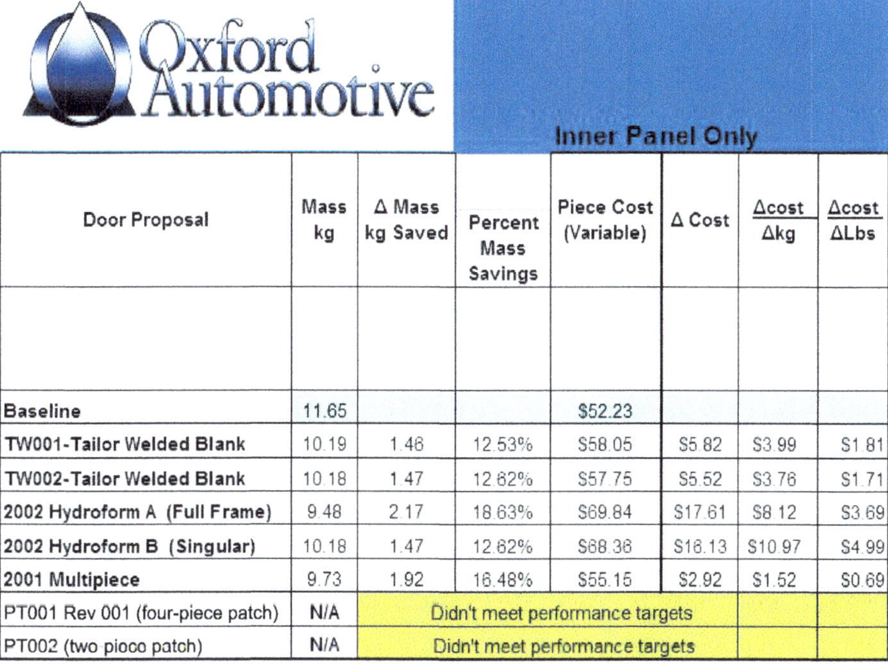

Door Proposal	Mass kg	Δ Mass kg Saved	Percent Mass Savings	Piece Cost (Variable)	Δ Cost	Δcost / Δkg	Δcost / ΔLbs
Baseline	11.65			$52.23			
TW001-Tailor Welded Blank	10.19	1.46	12.53%	$58.05	$5.82	$3.99	$1.81
TW002-Tailor Welded Blank	10.18	1.47	12.62%	$57.75	$5.52	$3.76	$1.71
2002 Hydroform A (Full Frame)	9.48	2.17	18.63%	$69.84	$17.61	$8.12	$3.69
2002 Hydroform B (Singular)	10.18	1.47	12.62%	$68.36	$16.13	$10.97	$4.99
2001 Multipiece	9.73	1.92	16.48%	$55.15	$2.92	$1.52	$0.69
PT001 Rev 001 (four-piece patch)	N/A	Didn't meet performance targets					
PT002 (two pioco patch)	N/A	Didn't meet performance targets					

Figure 3.51 Phase II performance data

At the end of Phases I and II, the following were concluded:

1. To achieve large weight decreases with low variable cost increases, the challenge would be to overcome the perceived dimensional problems associated with multipiece constructions.
2. The door inner, while it represents the largest piece of the door-in-white, presents large hurdles in terms of significant weight reduction at low cost. This is because the door inner is stiffness driven and cannot take advantage of AHSS, and the simple LWB door inner, which is already in production on roughly half of the front doors, is already architecturally robust. This was especially true with the donor door used for this study because it had a 3-to-1 thickness ratio of the front of the door versus the rear of the door. Most manufacturers are limiting this ratio to closer to 2-to-1.
3. Preliminary analysis of the door outer panel indicated that a reduction in thickness from .75 mm to .60 mm, while substituting AHSS for the production IF rephos material, could provide a 20% reduction in the outer without increasing cost. Also, the door outer is the second largest component at 40% of the door-in-white
4. It was decided that the focus of Phase III would be on the door outer, knowing that several technical hurdles needed to be overcome, including the following:
    a. Repeatable stamping of thin gauge AHSS would require establishing stamping guidelines.
    b. Surface quality requirements would have to be met, while simultaneously meeting all of the other door functional requirements, including dent resistance, door sag, stiffness, and so forth.

The objective of Phase III (Figure 3.52) therefore became to prove the viability of manufacturing door outers in dual-phase (DP) 500 or bake hardenable (BH) 250 steels at thicknesses of 0.6, 0.65, and 0.70 mm. Based on this goal, the following technical approach was used:

1. Get quotes from five companies to lead the project.
2. Using the same baseline door, do detailed performance and stampability analysis.
3. Build multiple doors with six different combinations of gauge (0.60 mm to 0.70 mm), grade (BH 250 and DP 500), and coating (electrogalvanized and galvannealed).

Figure 3.53 shows the results (i.e., CAE and testing) for our first stamping condition (i.e., no product or styling changes from the baseline). One trend shown in this table that was replicated in our other subsequent steps was that the CAE results were much more conservative than the actual stamping results. Based on the actual stamping trials, using Kirksite soft tools, there were four conditions in which the high-strength material stamped successfully. Most notable was the fact that the DP 500, EG at .625 mm stamped successfully. Also, the DP 500 EG at .70 mm, the DP 500 EG at .65 mm, and the BH 250 EG at .70 mm stamped successfully. A couple of the other panels had splits in the mirror pocket area.

Phase 3 Project Objectives

The main objective of the Auto/Steel Partnership (A/SP) Lightweight Closure Project Phase 3 (2004) project effort was to investigate the reduction of the door outer panel thickness using advanced high-strenght steel by utilizing different materials/thicknesses and coating variations. Materials considered for this project were bake hardenable and dual phase steels, with electro-galvanized (EG) and hot-dipped galvannealed (HDGA) coatings.

Figure 3.52 Phase III project objectives

Stampable Matrix (Baseline) Analytical vs. Physical

Stampable w/o Product Change (Baseline)									
	ANALYTICAL					PHYSICAL			
MAT'L	GAUGE	MASS Kg	PERCENT SAVINGS	WORST	TYP	ACTUAL GUAGE	ACTUAL SAVINGS	ACTUAL	
Hft 340 GA	0.74mm	5.74	%	N/A	N/A	0.76mm	MASS Kg %	YES	BASELINE PRODUCT
DP 500 EG	0.70mm	5.37	6.4	(M.P.)	(M.P.)	0.72mm	0.22 Kg 3.8%	YES	Showed a local marginal area in the mirror pocket but was not a significant concern for a non acceptance of the panel
DP 500 HDGA	0.70mm	5.37	6.4	(M.P.)	(M.P.)	0.69mm	N/A	NO	Panel split in the mirror pocket area
DP 500 EG	0.65mm	4.98	13.2	(M.F.)	(M.F.)	0.67mm	0.60 Kg 10.4%	YES	Panel was acceptable
DP 500 EG	0.625 mm	4.8	16.4	(M.F.)	(M.F.)	0.62mm	0.99 Kg 17.2%	YES	Panel was acceptable
DP 500 HDGA	0.60mm	4.6	19.8	(M.F.)	(M.F.)	0.60mm	N/A	NO	Panel split in the mirror pocket area
BH250 EG	0.70mm	5.37	6.4	(M.F.)	(M.F.)	0.68mm	0.52 Kg 9.0%	YES	Panel was acceptable
BH250 EG	0.65mm	4.98	13.2	(M.F.)	(M.F.)	0.64mm	0.83 Kg 14.4%	N/A	Panel was not run due to timing issues on delivery of material (see revision 11A)
		1. Revision in window at rear lower area in trimmed out surface							
LEGEND		2. tear drop cut out in window opening							
(MF)	Failure in Mirror Pocket (analytical)								
(MP)	Mirror pocket close to acceptable result								
(MP)	Acceptable Results								

Figure 3.53 Stampability matrix using the production exterior surface

We then tried to get the two failed DP conditions to stamp by allowing a slight product change that would not affect the styling of the door (Figure 3.54). However, we did not improve the stamping condition by modifying the die for a minor product change.

Stampable Matrix (10A) Analytical vs. Physical

Stampable with Minor Product Change w/o Styling Change						(Rev. 10A)	
	ANALYTICAL			ANALYTICAL		PHYSICAL	
MAT'L	GAUGE	MASS Kg	PERCENT SAVINGS	WORST	TYP	ACTUAL	
DP 500 EG	0.70mm	5.37	6.4	(M.P.)	(M.P.)	—	Panel was acceptable in the baseline iteration
DP 500 HDGA	0.70mm	5.37	6.4	(M.P.)	(M.P.)	NO	Panel failed in the mirror pocket area
DP 500 EG	0.65mm	4.98	13.2	(M.P.)	(M.P.)	—	Panel was acceptable in the baseline iteration
DP 500 EG	0.625mm	4.8	16.4	(M.P.)	(M.P.)	—	Panel was acceptable in the baseline iteration
DP 500 HDGA	0.60mm	4.6	19.8	(M.P.)	(M.P.)	NO	Panel failed in the mirror pocket area
BH250 EG	0.70mm	5.37	6.4	(M.P.)	(M.P.)	—	Panel was acceptable in the baseline iteration
BH250 EG	0.65mm	4.98	13.2	(M.P.)	(M.P.)	N/A	Panel was not run due to timing issues on delivery of material (see revision 11A)
BH250	0.60mm	4.6	19.8	(M.P.)	(M.P.)	N/A	Material was not available
				1. Revision in window at rear lower area in trimmed out surface			
LEGEND				2.Tear drop cut out in window opening			
(MP)	Failure in mirror pocket (analytical)			3.Mirror pocket depth moved outboard 5.0 mm			
(MP)	Mirror pocket close to acceptable result						
(MP)	Acceptable results						

Figure 3.54 Stampability matrix with a minor product change

We then decided to modify the die for a minor product change and a minor styling change with the hope of getting one or more panels to stamp (Figure 3.55). Under this scenario, we were able to get one of the other panels to stamp (i.e., the BH 250 EG at .65 mm).

Stampable Matrix (11A) Analytical vs. Physical

Stampable with Minor Product Change and Minor Styling Change						(Rev. 11A)	
	ANALYTICAL			ANALYTICAL		PHYSICAL	
MAT'L	GAUGE	MASS Kg	PERCENT SAVINGS	WORST	TYP	ACTUAL	
DP 500 EG	0.70mm	5.37	6.4	(M.P.)	YES	—	Panel was acceptable in the baseline iteration
DP 500 HDGA	0.70mm	5.37	6.4	(M.P.)	YES	NO	Panel Failed in the mirror pocket area
DP 500 EG	0.65mm	4.98	13.2	(M.P.)	(M.P.)	—	Panel was acceptable in the baseline iteration
DP 500 EG	0.625mm	4.8	16.4	(M.P.)	(M.P.)	—	Panel was acceptable in the baseline iteration
DP 500 HDGA	0.60mm	4.6	19.8	(M.P.)	(M.P.)	NO	Panel Failed in the mirror pocket area
BH250 EG	0.70mm	5.37	6.4	(M.P.)	(M.P.)	—	Panel was acceptable in the baseline iteration
BH250 EG	0.65mm	4.98	13.2	(M.P.)	(M.P.)	YES	Panel was acceptable
BH250	0.60mm	4.6	19.8	(M.P.)	(M.P.)	N/A	Material was not available
				1. Revision in window at rear lower area in trimmed out surface			
				2.Tear drop cut out in window opening			
LEGEND				3.Mirror pocket depth moved outboard 5.0 mm			
(MP)	Failure in mirror pocket (analytical)			4. Revision surface to angle off the radius of the character line to the pocket at bottom side of mirror flag in localized area			
(MP)	Mirror pocket close to acceptable result						
(MP)	Acceptable results						

Figure 3.55 Stampability matrix with minor product change and minor styling change

The Lightweight Closures Project–Phase III was successful in achieving mass reductions up to 17.2% on the door-outer panel by demonstrating that the baseline 0.74 mm IF rephos material could be changed to the following:

1. BH 250 at 0.64 mm with a slight styling concession.
2. DP 500 at gauges as low as 0.62 mm within the baseline die configuration.
3. Testing results showed that the DP steels performed much better for dent resistance than the baseline material.

In terms of the whole Lightweight Closure Project:

1. The technologies, which were investigated within the A/SP Lightweight Closures Project–Phases I through III, demonstrated affordable (less than $0.70 per pound saved) methods of achieving 15 to 20% weight reductions.
2. The materials, which were investigated with the Lightweight Closures Project–Phase III, could provide better dent resistance at thinner gauges.
3. The design and draw guidelines developed within the Lightweight Closures Project–Phase III, would be generally applicable to exterior panels with higher strength material.

This project was also useful in identifying areas that would require future work:

4. Doors can be designed to provide a better tie-in with the rest of the body, thereby providing a more efficient body.
5. Trimming, flanging, and hemming were not directly addressed in the Lightweight Closures Project. These constraints could be very significant in the successful implementation of higher strength door outers.
6. More accurate methods for springback prediction need to be developed.
7. General die design, material, lube, and maintenance guidelines need to be further developed for addressing these higher strength steels.

Based on this project, there are opportunities to increase the strength and reduce the thickness of current exterior closure panels.

3.5 Reference

3-1. EDAG Inc., 2012, *Mass Reduction for Light-Duty Vehicles for Model Years 2017–2025*, DOT HS 811 666, U.S. Department of Transportation, National Highway Traffic Safety Administration.

Chapter 4
Comparison of Advanced High-Strength Steels with Alternative Materials

4.1 Material Comparison Overview

In this chapter, advanced high-strength steels (AHSS) will be compared with other materials. The learning objectives will be the following:

1. To understand the advantages and disadvantages of using steel versus aluminum and other alternative materials.
2. To understand the cost advantages of using steel for weight reduction.
3. To be able to quantify the mechanical benefits of steel versus alternative materials.

Figure 4.1 shows the various structural materials for automotive applications on a cost versus stiffness/strength scale. This plot is not meant to be quantitative, but is meant to show roughly how the different materials "stack up." Aluminum is more than four times the cost of mild steel and offers some weight advantages versus steel, but it has one-third the modulus and is not as strong as the higher strength steels. Because it is a close competitor with steel for automotive applications, this chapter will focus on comparisons of aluminum versus steel. Aluminum has been making strong gains versus steel for closure applications because (as you will see later) aluminum has more weight-saving potential for closures versus body-in-white applications.

Magnesium has been used for various automotive structural parts (e.g., instrument panels), but it lacks enough strength and ductility to be used in major automotive structural areas. However, because it can be used in die castings down to about 2 mm, it competes with some of the parts that have been converted to aluminum. For instance, magnesium has been used for the inner panel for some closures (e.g., hatchbacks on sport utility vehicles).

Conventional composites such as sheet molding compound (SMC) lack the stiffness and strength to compete with the metals for weight-saving potential. A common application of conventional composites is for exterior panels (e.g., fenders, hoods, and pickup box sides), where the shapes discourage the use of metals from a stamping point of view. Carbon fiber composites are still much too costly for major automotive applications.

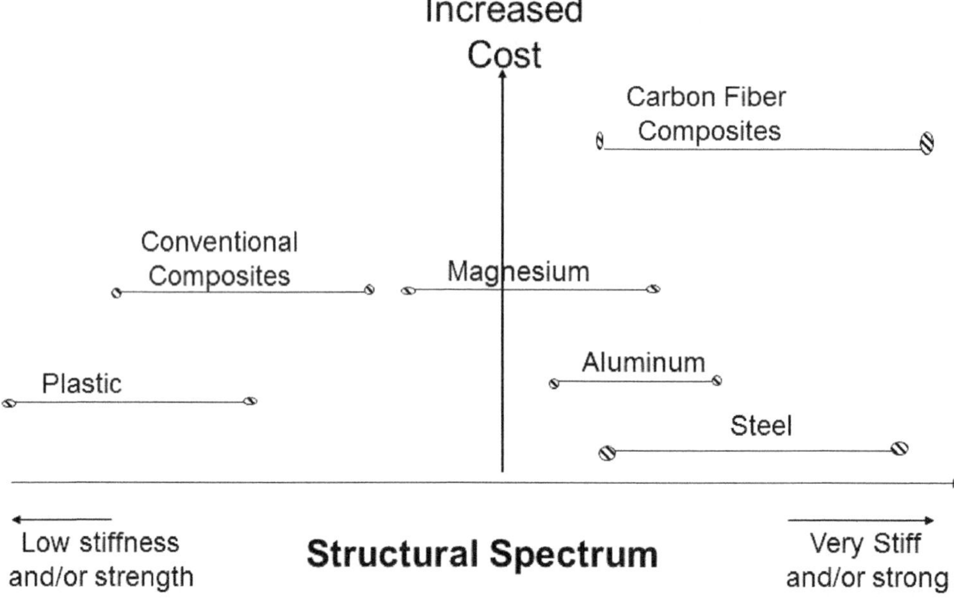

Figure 4.1 Automotive material plot

4.2 Elementary Structural Mechanics

In this and the next two sections, steel will be compared to aluminum for weight saving potential from a basic strength-of-materials perspective. First, weight differences for steel and aluminum for a simple plate will be compared (Figure 4.2).

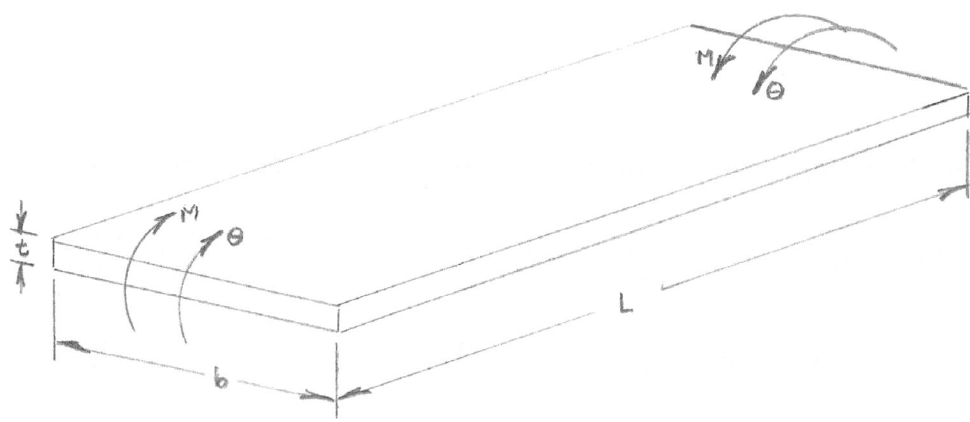

Figure 4.2 Plate under bending load

If it is given that:

M = Moment applied to plate in lb.-in.

Θ = Angle of bending of plate in radians

L = Length of plate in inches

b = Width of plate in inches

ρ = Density of material in lb./in.^{3}

Density of aluminum = 0.1 lb./in.^{3}

Density of steel = 0.28 lb./in.^{3}

E = Modulus of elasticity in. psi.

Modulus of elasticity for aluminum = 10×10^6 psi

Modulus of elasticity for steel = 30×10^6 psi

Bending stiffness = $(E \times b \times t^3)/(12 \times L)$ [4-1]

Weight = $b \times t \times L \times \rho$

Problem: Find the ratio of the weight of the steel plate to the aluminum plate if the bending stiffness, b and L are the same. Hint: $^{1/3}\sqrt{3} \approx 1.4$

The answer is: $W_s/W_a \approx 2$. Another way of saying this is that if a steel plate is converted to aluminum, 50% of the weight can be saved for the same stiffness. In this case the aluminum plate will be 1.4 times the thickness of the steel plate. This will also be the case if the plate is under a twisting load.

The next exercise will be for a thin-walled tube (Figure 4.3) in torsion.

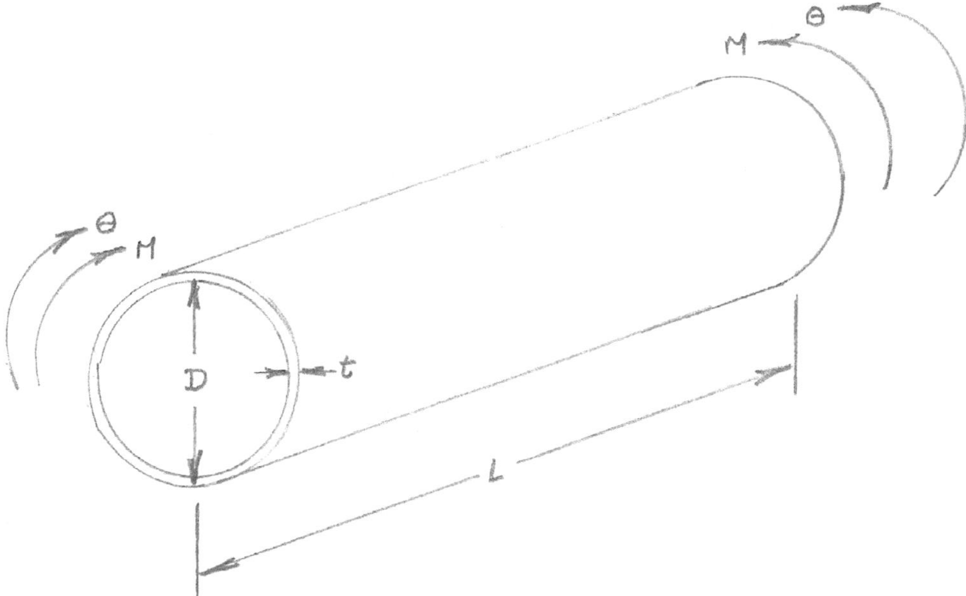

Figure 4.3 Thin-walled tube in torsion

If we are given that:

M = Moment twisting the tube in lb.-in.

Θ = Angle of twist of the tube in radians

L = Length of the tube in inches

D = Diameter of the tube in inches where R = D/2

ρ = Density of material in lb./in.^{3}
Density of aluminum = 0.1 lb./in.^{3}
Density of steel = 0.28 lb./in.^{3}

G = Shear modulus of elasticity in. psi.
Shear modulus of elasticity for aluminum = 4×10^6 psi
Shear modulus of elasticity for steel = 12×10^6 psi

Torsional stiffness = $(2 \times \pi \times G \times t \times R^3)/L$ [4-1]

Weight = $2 \times \pi \times R \times t \times \rho \times L$

Problem: Find the ratio of the weight of a steel tube to an aluminum tube if the torsional stiffness, R and L, are the same.

Then the answer is: $W_s/W_a \approx 1$. Another way of saying this is that if a steel tube is converted to aluminum, no weight will be saved for the same torsional stiffness. In this case, the aluminum tube will have a wall thickness 2.8 times the thickness of the steel plate. This is also true if the tube is in bending or tension, or for a Euler buckling load case.

Since an automobile body can be considered as a combination of plates and tubes, you might surmise that from a stiffness perspective, the weight savings for a conversion to aluminum should be somewhere between 0 and 50%. However, the tubes of the car are usually rectangular in shape, formed by the intersection of two pieces of sheet metal. As the sides of the rectangle get more unequal (i.e., the rectangle gets closer to a plate), the weight savings get closer to 50% for a conversion to aluminum. Also, as the height of the rectangle, in the plane of the load, gets very large compared to the width, the weight savings also gets close to 50%. Another aspect of the rectangular cross sections that make up the major structural members of the automobile is that the seams along the length of the beam are typically not continuously welded. Even for a round tube, with an open weld seam, there is a 50% weight reduction for a conversion to aluminum. However, the beams of the car have a 6 mm weldment at approximately 50 mm apart along the length of the beam. This means that even for a round beam the weight advantage for a spot welded beam would be somewhere between 0 and 50% with a conversion to aluminum.

This discussion would lead one to believe that for traditional automobile architectures, using mild steels, a conversion to aluminum would save close to 50% in weight if stiffness were the only concern. This would give some justification for the conclusions in the U.S. government's Program for a New Generation of Vehicles (PNGV) program of the mid-1990s, where automobiles were primarily stiffness driven. However, because of the latest generation of safety regulations, automobiles and light trucks are primarily safety driven, where safety is the attribute that dictates whether weight needs to be added to the vehicle during the vehicle development cycle. Where aluminum may have some advantages for complex buckling modes because of the greater thicknesses, new grades of AHSS have given steel an advantage from a strength perspective. Also, improved architectures (e.g., hydroformed tubes, structural adhesives, and laser welding) have also given steel an advantage, in that the beams of the car are approaching continuously joined sections.

This section presented the fundamental engineering mechanics theory for steel to aluminum conversion and gave a qualitative idea of weight savings potential for steel to aluminum conversion. The following few sections will expand on this theory and give a more quantitative idea of the relative weights of aluminum and steel vehicles.

4.3 Generalization of Elementary Theory

In this section, the theory developed in the previous section will be generalized. This theory was an outgrowth of the Improved Materials & Powertrain Concepts for 21st Century Trucks (IMPACT) Phase I project, which was led by Dr. Richard Patton (a professor at Mississippi State University), Dr. Glen Prater (now chairman of the Department of Mechanical Engineering) at the University of Louisville, and the author. The specific theory addressed here is the Lambda (λ) factor theory, which was invented by Dr. Patton. The lambda factor methodology was developed as a unified theory for calculating the weight benefits of converting from one material to another (e.g., converting from steel to aluminum). Though there were several previous studies on the weight comparison of steel and aluminum (e.g., PNGV) done before IMPACT was begun, no one really developed an in-depth understanding of specifically why aluminum could reduce weight to the stated amount. More generally, no one developed a rationale for changing the architecture of a vehicle so that steel could perform better versus aluminum in weight studies. This was of concern to the U.S. army, because they were looking at the next generation of HUMVEE and were trying to develop a rationale as to whether the body of the HUMVEE should be steel or aluminum, and to Ford, because they were looking at possible steel alternatives for their next generation of pickup trucks. A possible source of generalization stems from the plate versus tube comparison from the preceding section (Figure 4.4). Since the bending stiffness of the plate is proportional to the inertia of the cross section of the plate and the torsional stiffness of the tube is proportional to the polar moment of inertia of the tube, notice that the inertias are proportional to the thickness cubed for the plate and proportional to the first power of the thickness for the tube.

In general, steel and aluminum have approximately the same specific stiffness, defined as the modulus of elasticity divided by the density. Generally speaking, the response to an increase in metal thickness is the most important criterion for weight savings via material substitution, where a linear response favors steel, which at equal weight has much reduced cost, and a non-linear response favors aluminum. In the case of the plate, the stiffness is proportional to the thickness cubed, and in the case of the tube the response is related to the thickness to the first power. For our theory, λ is assigned the value of the power of the thickness in the equation for the response. So for our example, λ equal to three means that the conversion from steel to aluminum will give a 50% reduction in weight for the conversion from steel to aluminum; and for λ equal to one, the weights will be the same (Figure 4.5). The final formula for weight saved for steel to aluminum conversion is shown in Figure 4.6. By the way, similar relationships can be derived when comparing any two metals that have the same specific stiffness.

The Flat Plate vs. the Tube

- Flat plate has bending stiffness dependent on inertia
- $I=1/12*b*t^3$
- $\lambda = 3$
- Thin-walled tube has torsional stiffness dependent on polar moment of inertia
- $J \sim 2*\pi*r^3*t^1$
- $\lambda = 1$

Figure 4.4 Flat plate versus tube comparison

Weight savings by switching from steel to aluminum – flat plate and tube

- The bending stiffness increased as the cube of the thickness for the flat plate
- Weight savings for the flat plate are:
  - $(\rho_{al}/\rho_{st})*(E_{st}/E_{al})^{1/3} \sim 50\%$ ($\lambda = 3$)
- The torsional stiffness increased linearly with the thickness of the tube
- Weight savings for the tube are:
  - $(\rho_{al}/\rho_{st})*(E_{st}/E_{al})^{1/1} \sim 0\%$ ($\lambda = 1$)
- Weight savings are completely dependent on λ!

Figure 4.5 Steel to aluminum conversion factors

Beam thickness exponent, λ

- Determines weight saved by going to aluminum from steel
- Formula is:
  - $m_{ratio} = (\rho_{al}/\rho_{steel})^*(E_{steel}/E_{al})^{1/\lambda}$
  - % Weight saved = $(1 - m_{ratio})^*100$

Figure 4.6 Weight saved as a function of λ

From elementary structural mechanics, we know that a hollow tube is a more structurally efficient way to transfer torsional or bending loads than a flat plate (Figure 4.7). A corrugated plate would be somewhere in between those two. It can be inferred from this fact and from what has been said previously that the more structurally efficient the structure, the less advantage aluminum would have versus steel from a weight perspective. So the goal of automotive structural engineers should be to achieve more structural efficiency if they want to stay in low-cost steel.

Beams in bending

- A flat plate can be considered as the extreme of a continuum of forms.
- A corrugated surface is more efficient than a flat plate in bending. It is a transitional form between flat plates and hollow beams. Other forms include sandwich materials.
- A hollow beam is more efficient than a corrugated surface.

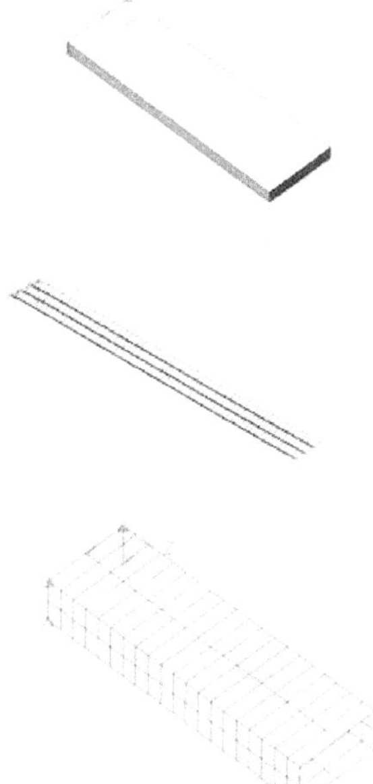

Figure 4.7 Beams in bending

Another way of summarizing what we have covered so far is to look at the plot of the hollow beam thickness exponent versus 2 times the thickness (t) divided by the height between the top and bottom flanges of the beam (Figure 4.8). The height (h) is zero for a flat plate, and h equals the overall height minus 2 times t for other forms. For a given width of beam, as the height goes to infinity, λ approaches 1, which means that there is no advantage for converting steel to aluminum as far as weight is concerned. As h approaches zero, λ approaches 3, which means that the aluminum plate would weigh half as much as the steel plate. A corrugated plate would be somewhere in between. For other structural shapes, other than simple beams, λ would typically be somewhere between 1 and 3.

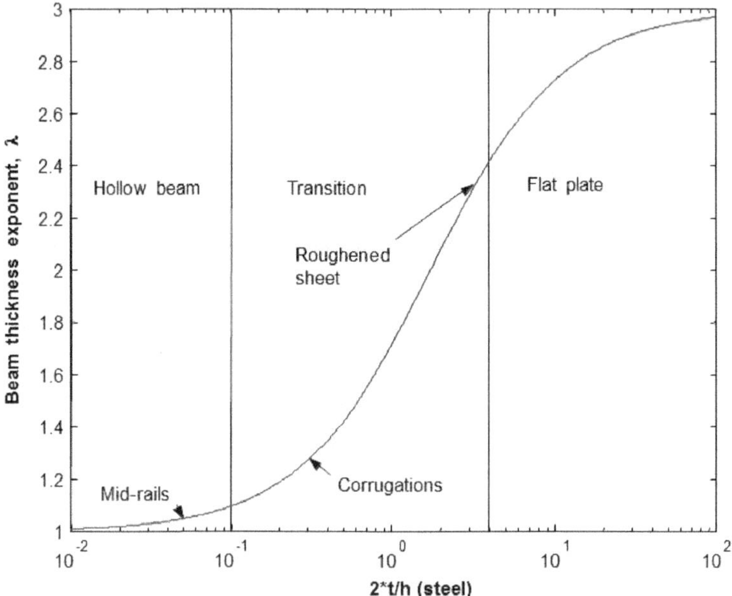

Figure 4.8 Beam thickness exponent

Therefore the plot shown in Figure 4.8 can be redrawn to represent % weight saved in converting to aluminum versus the same abscissa (Figure 4.9).

There is also a way to compute λ for a beam without knowing its moment of inertia if you have a finite element model of the beam. Two runs of the finite-element model are needed using two different thicknesses while subjected to the same force or moment. Then compute the displacements for the two runs and compute λ from the displacements and thicknesses as shown in Figure 4.10.

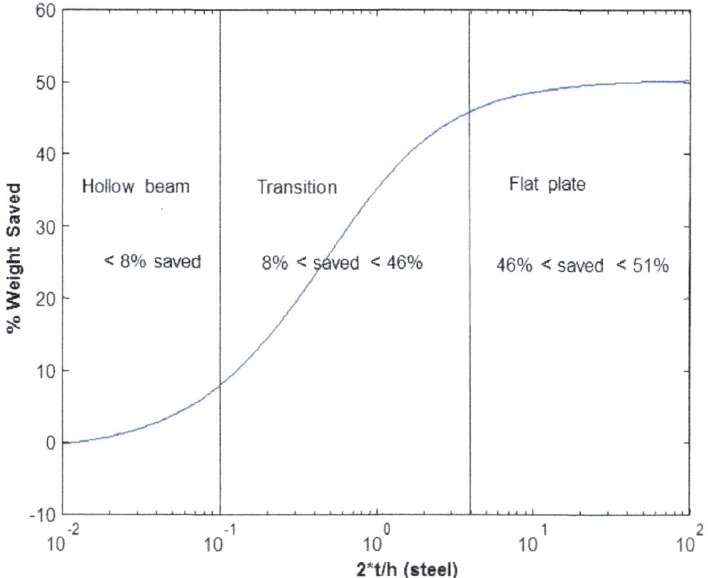

Figure 4.9 Aluminum versus steel weight savings: geometric beam effects

Summary of λ and Beam Geometric Effects

- λ can be used to categorize different structures
- Different beam geometries result in different λs
- Generally speaking, the more efficient the structure, the lower λ will be
- λ can be computed using finite elements, by making two runs, identical except for changing the thickness of the elements, from the formula:
  - λ = log(d1/d2)/log(t1/t2), where
    - d = representative displacement for runs 1 & 2
    - t = thickness for runs 1 & 2
  - In these runs, the material should be kept the same

Figure 4.10 Computing lambda from a finite-element model

Actually, this same technique can be used to compute λ for any structure that you have a finite-element model of, if the structure being evaluated has nearly the same thickness for all of its parts. For instance, suppose you had a model of a car body that has thicknesses of nearly 1 mm for each of its parts. You could set all the thicknesses to 1 mm and make a run for a torsional load. A similar run could be made for 2 mm panels and λ computed. This should give a rough idea of how much weight could be saved by converting to aluminum. However, this would only give an idea of how much weight could be saved if the only load on the car was static torsion. λ could also be computed for each load of a representative load set, and λ could be computed for each load case. However, no one has reported using this concept for a whole car or even a car body. This might be a useful study for the future.

This theory will be extended for other structural entities other than simple beams. We will first examine curved beams, which also can be treated using λ. This includes all the joints in the vehicle, as well as curved beams. A T-joint can be considered to be a double curved beam with two inside flanges. Examine the curved joint/beam shown in Figure 4.11; a finite-element model of the beam is depicted.

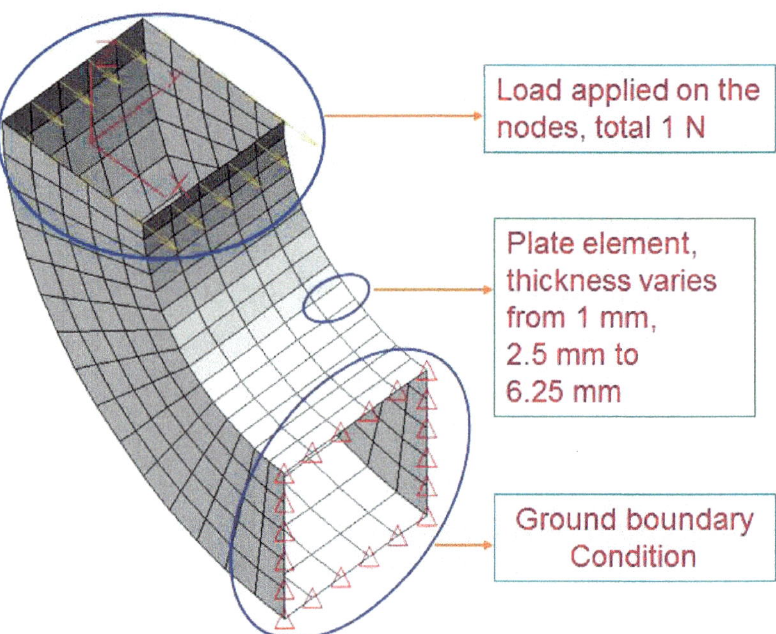

Figure 4.11 Curved beam bending model

The nodes at the right-hand open section are grounded and a load of a total of 1 Newton is applied at the open section at the top. Three runs are made at 1 mm, 2.5 mm, and 6.25 mm. Figure 4.12 shows the results in terms of a stress tensor superimposed on the three models. The compressive stresses increase on the inside top flange considerably, and the tensile stresses increase on the outside bottom curved flange when the thickness is increased to 6.25 mm. The stresses decrease on the side flanges. This would indicate a much more efficient use of material, and appropriately λ drops to 1.1.

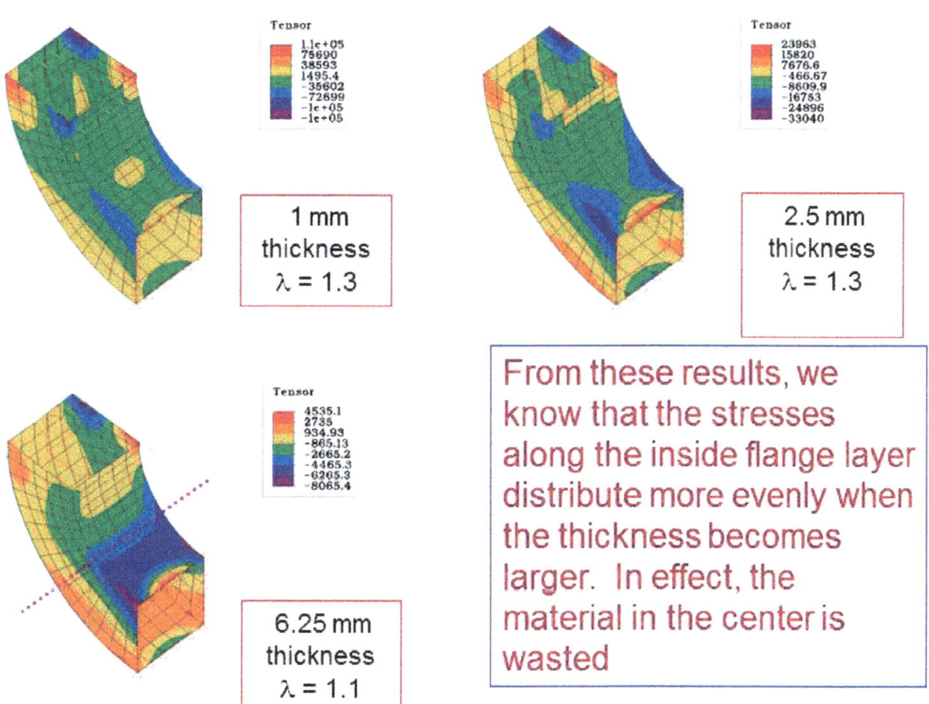

Figure 4.12 Curved-beam bending results

Now it is appropriate to examine actual vehicle joints. If a body or frame can be considered to be a combination of beams, joints, and plates, experience shows that improving joint efficiency will have a larger effect on improving overall vehicle stiffness. In Figure 4.13, an A-pillar to roof joint from a vehicle is shown, which was benchmarked for the IMPACT study. Fore-and-aft and lateral loading conditions are applied to this joint. λ is computed, using a procedure similar to that shown in Figure 4.10, for each loading condition, and it comes out to be 1.40 for the first loading condition and 1.51 for the second. These results would indicate that the weight savings, using an estimate from Figures 4.8 and 4.9, would be somewhere between 25 and 35% with a conversion to aluminum.

A Pillar/Roof Joint

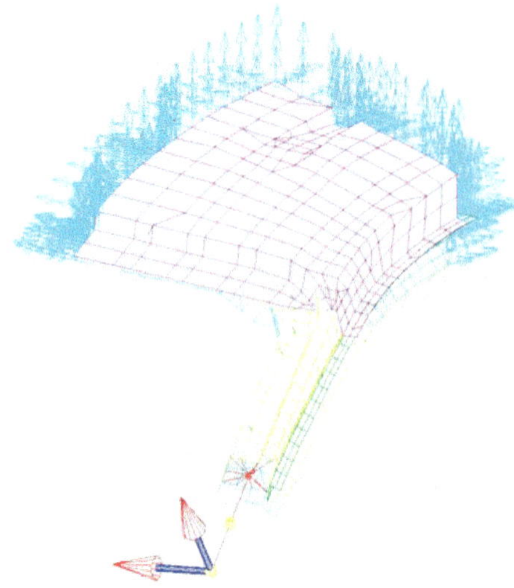

Force	Initial Displacement	Lambda
F/A	1.8E-4 mm	1.40
I/O	9.24E-4 mm	1.51

Figure 4.13 A-pillar to roof joint

Similarly, some other joints from the same vehicle were analyzed and the results are shown in Figure 4.14. Since the weight save with a conversion to aluminum was estimated from the generalized beam result, the stiffness was slightly off the baseline steel result when the joint was converted to aluminum. This error is shown in the right-hand column of Figure 4.14.

Joint Analysis Results

Joint	Load	λ	% Weight Saved	Stiffness Error
B/Roof	F/A	1.58	32%	-2.2%
	I/O	2.00	41%	-0.4%
B/Rocker	F/A	1.39	25%	2.3%
	I/O	1.54	30%	2.9%
Hinge/ Rocker	F/A	1.33	22%	0.4%
	I/O	1.46	28%	-7.7%

Figure 4.14 Joint analysis results

Beams and joints in the vehicle can be made more structurally efficient by adding reinforcements or rigidizing the beam or joint through better joining techniques (e.g., using structural adhesives, laser welding, and so on). Figure 4.15 shows a joint with reinforcement added. This reinforcement consists of a plate inserted in the center of the joint and welded between the weld flanges as shown in yellow in the figure. The results with and without the reinforcement are shown in Figure 4.16.

B-Pillar to Rocker – Original and Reinforced

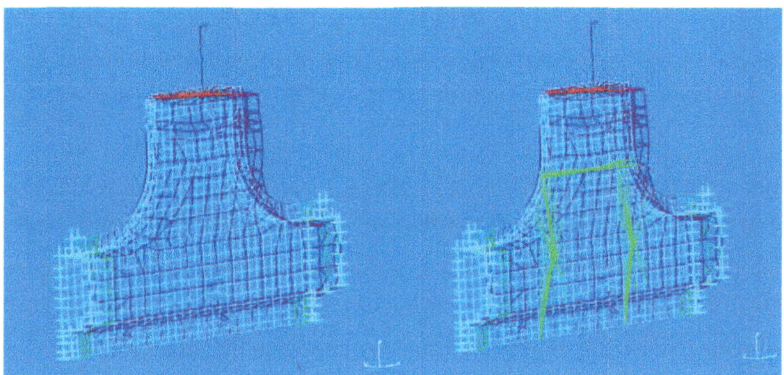

Figure 4.15 B-pillar to rocker reinforcement

Reinforcement Results

Inboard-outboard Loading	λ	% Von-Mises Stress Reduction	% Stiffness Increase
Original Model	1.5474		
Model with Reinforcement	1.1517	53.6%	315%

Fore-aft Loading	λ	% Von-Mises Stress Reduction	% Stiffness Increase
Original Model	1.3893		
Model with Reinforcement	1.144	49.0%	293%

Figure 4.16 Reinforcement results

For both loading conditions, λ drops appreciably for the reinforced joint, the stress drops, and the stiffness increases. The following are overall conclusions from the joint analysis:

1. The response of a structure to changes in wall thickness can be used to predict the weight savings from material substitution.
2. Lambda (λ) is a very good predictor of weight savings for equivalent stiffness.
3. Of the joints benchmarked, the average λ was 1.51, which gave an average weight savings of 30% for steel to aluminum material substitution.
4. Variations in geometry between joints cause changes in λ for the different joints.
5. Lambda (λ) for a joint can be reduced through redesign and reinforcement.

Though the λ factor theory appears to be a useful way of generalizing simple engineering mechanics calculations, there is still significant work to be done to fully take advantage of this theory:

1. Investigate the possibility of defining a global λ (i.e., full body or full vehicle) based on forced dynamic response, including compliance frequency response and crash.
2. Develop the analytical background (calculation methodology, weight savings predictions) for a global λ based on other benchmarking metrics (e.g., diagonal distortions).
3. Use simplified analytical tools to continue to investigate the influence of thin-wall effects on λ behavior and interpretation.
4. Conduct comparative global gauge sensitivity studies for a variety of passenger car and truck bodies.
5. Investigate the use of global λ curves to assess the effectiveness of local architectural changes.

4.4 Recent Comparisons of Advanced High-Strength Steels and Aluminum

There have been several studies done at either the full body or the full vehicle level that have been used to compare weight reduction potential in steel or aluminum. Most of these studies have been initiated by material suppliers or original equipment manufacturers (OEM). This has led to some controversy as to whether bias has come into play in these studies. Recently, four studies have been sponsored by the California Air Resource Board (CARB), the Environmental Protection Agency, and the National Highway Traffic and Safety Administration (NHTSA); and the studies were conducted by very competent vehicle design houses (i.e., Lotus Engineering, FEV, and EDAG):

1. 2010: CARB/Lotus Engineering–Initial Paper Study on Toyota Venza (Phase 1)
2. 2012: EPA/FEV (Phase 2) 2010 midsize CUV

3. 2012: NHTSA/EDAG Study on Honda Accord
4. 2012: CARB (Phase 2) 2009 Midsize CUV high-development vehicle

A comparison of these studies and a detailed analysis of the second study can be found at www.autosteel.org [4-2]. For the purpose of this book, the focus will be on the third study in the list because it used a highly developed architecture (i.e., 2011 Honda Accord); it used a mid-sized sedan, which represents the highest worldwide volumes and which was used for the PNGV vehicle studies in the early 1990s; the critical data, used and developed in the study, are in the public domain; and this study has been used in some very recent studies that point to the future direction of AHSS. Also, the company (EDAG) that led this study also led the Future Steel Vehicle Study [4-3], which developed some of the critical data used in the Honda Accord study. The material used in this section will be drawn from the talk given by the Program Manager, Harjinder Singh, from EDAG at the 2012 Great Designs in Steel (GDIS) Conference [4-4], and the final report, which was published by NHTSA [4-5]. This study had the specific objectives to identify light weighting technologies for 2020 model year vehicles, to consider costs +/–10% of current vehicles; the new lightweight vehicle (LWV) should have the same performance and functionality as the base vehicle; all recommended technologies to be suitable for 200,000 annual production and 1 million vehicles over 5 years; and the study had to deliver a detailed computer-aided engineering (CAE) model to NHTSA for future safety-related work. The materials considered for this study were steel, aluminum, magnesium, and plastics, including sheet molding compound (SMC). The steel materials considered are shown in Figure 4.17 along with the properties. These properties were obtained by EDAG from the major world steel companies during the Future Steel Vehicle (FSV) project. Super forming of aluminum, stamping of and warm forming of magnesium, and composites, except for SMC, were considered to be not ready for high-volume 2020 production. In terms of manufacturing assembly operations, spot welding, laser welding, mig welding, laser brazing, adhesive bonding, and mechanical fasteners were considered. Though this study was a full vehicle study, only sheet metal applications for the body structure, closures, and chassis will be considered here.

The four body structure options that were considered for the study are shown in Figure 4.18. The body structure for this study excludes closures and nonstructural items. The first option was an AHSS intensive design. The second option was an AHSS design with an aluminum roof panel, and a PPG floor and package tray. The third option was an aluminum intensive design, and the fourth option was a composite design.

The end result of the body structure study was that 22.2% of the weight could be removed from the baseline with the AHSS intensive design, whereas 24.5% could be taken out with option 2, and 35% with option 3. Though the composite solution yielded a 50% reduction, this solution was deemed to be inappropriate given the ground rules for the study. Though we will not be addressing costs until the next section, the AHSS intensive design appears to be the high-value alternative, given that option 2 gave only a 2.3% improvement beyond the 22.2% of option 1, and option 3 gave only a 12.8% improvement over option 1.

WorldAutoSteel — FSV Materials Portfolio — Updated March 4, 2011

Item #	Steel Grade	Thickness (mm) Min t	Thickness (mm) Max t	Gage Length	YS (MPa) Min	YS (MPa) Typical	UTS (MPa) Min	UTS (MPa) Typical	Tot EL (%) Typical	N-value Typical	Modulus of Elasticity (MPa)	Fatigue Strength Coeff (MPa) *	K Value (MPa)
1	Mild 140/270	0.35	4.60	A50	140	150	270	300	42-48	0.24	21.0×10^4	645	541
2	BH 210/340	0.45	3.40	A50	210	230	340	350	35-41	0.21	21.0×10^4	695	582
3	BH 260/370	0.45	2.80	A50	260	275	370	390	32-36	0.18	21.0×10^4	735	550
4	BH 280/400	0.45	2.80	A50	280	325	400	420	30-34	0.16	21.0×10^4	765	690
5	IF 260/410	0.40	2.30	A50	260	280	410	420	34-48	0.20	21.0×10^4	765	690
6	IF 300/420	0.50	2.50	A50	300	320	420	430	29-36	0.19	21.0×10^4	775	759
7	FB 330/450	1.60	5.00	A80	330	380	450	490	29-33	0.17	21.0×10^4	835	778
8	HSLA 350/450	0.50	5.00	A80	350	360	450	470	23-27	0.16	21.0×10^4	815	807
9	DP 300/500	0.50	2.50	A80	300	345	500	520	30-34	0.18	21.0×10^4	865	762
10	HSLA 420/500	0.60	5.00	A50	420	430	500	530	22-26	0.14	21.0×10^4	875	827
11	FB 450/600	1.40	6.00	A80	450	530	560	605	18-26	0.15	21.0×10^4	950	921
12	HSLA 490/600	0.60	5.00	A50	490	510	600	630	20-25	0.13	21.0×10^4	975	952
13	DP 350/600	0.60	5.00	A80	350	385	600	640	24-30	0.17	21.0×10^4	985	976
14	TRIP 350/600	0.60	4.00	A50	350	400	600	630	29-33	0.25	21.0×10^4	975	952
15	SF 570/640	2.90	5.00	A50M	570	600	640	660	20-24	0.08	21.0×10^4	1005	989
16	HSLA 550/650	0.60	5.00	A50	550	585	650	675	19-23	0.12	21.0×10^4	1020	1009
17	TRIP 400/700	0.60	4.00	A80	400	420	700	730	24-28	0.24	21.0×10^4	1075	1077
18	SF 600/780	2.00	5.00	A50	600	650	780	830	16-20	0.07	21.0×10^4	1175	1201
19	HSLA 700/780	2.00	5.00	A50	700	750	780	830	15-20	0.07	21.0×10^4	1175	1200
20	CP 500/800	0.80	4.00	A80	500	520	800	815	10-14	0.13	21.0×10^4	1160	1183
21	DP 500/800	0.60	4.00	A50	500	520	800	835	14-20	0.14	21.0×10^4	1180	1303
22	TRIP 450/800	0.60	2.20	A80	450	550	800	825	26-32	0.24	21.0×10^4	1170	1690
23	CP 600/900	1.00	4.00	A80	600	615	900	910	14-16	0.14	21.0×10^4	1255	1301
24	CP 750/900	1.60	4.00	A80	750	760	900	910	14-16	0.13	21.0×10^4	1255	1401
25	TRIP 600/980	0.90	2.00	A50	550	650	980	990	15-17	0.13	21.0×10^4	1335	1301
26	TWIP 500/980	0.80	2.00	A50M	500	550	980	990	50-60	0.40	21.0×10^4	1335	1401
27	DP 700/1000	0.60	2.30	A50	700	720	1000	1030	12-17	0.12	21.0×10^4	1375	1521
28	CP 800/1000	0.80	3.00	A80	800	845	1000	1005	8-13	0.11	21.0×10^4	1350	1678
29	DP 800/1180	1.00	2.00	A50	800	880	1180	1235	10-14	0.11	21.0×10^4	1555	1700
30	MS 950/1200	0.50	3.20	A50M	950	960	1200	1250	5-7	0.07	21.0×10^4	1595	1678
31	CP 1000/1200	0.80	2.30	A80	1000	1020	1200	1230	8-10	0.10	21.0×10^4	1575	1700
32	DP1150/1270	0.60	2.00	A50M	1150	1160	1270	1275	8-10	0.10	21.0×10^4	1620	1751
33	MS 1150/1400	0.50	2.00	A50	1150	1200	1400	1420	4-7	0.06	21.0×10^4	1765	1937
34	CP 1050/1470	1.00	2.00	A50M	1050	1060	1470	1495	7-9	0.06	21.0×10^4	1840	2030
35	HF 1050/1500												
	Conventional Forming	0.60	4.50	A80	340	380	480	500	23-27	0.16	21.0×10^4	845	790
	Heat Treated after forming	0.60	4.50	A80	1050	1220	1500	1600	5-7	0.06	21.0×10^4	1945	2161
36	MS 1250/1500	0.50	2.00	A50M	1250	1265	1500	1520	3-6	0.05	21.0×10^4	1865	2021

Figure 4.17 FSV steel sheet metal properties

Options	Body Mass-kg	% Reduc. from Base	Mass Savings-kg
Baseline	328	0	0
AHSS Body	255.2	22.2	72.8
AHSS+ALU+PPG	247.5	24.5	80.5
Aluminum	213.2	35	114.8
Composite	164	50	164

Figure 4.18 Body structure options

As an example of the closure analysis, Figure 4.19 shows the results of the three alternatives considered for the front doors (i.e., AHSS intensive, aluminum intensive, and an aluminum/magnesium hybrid). In this case, the AHSS design gave only a 15% improvement over the baseline, which is consistent with the Auto/Steel Partnership (A/SP) Door Project. The aluminum intensive design gave a 48% improvement, which is due to the fact that a door is essentially two flat plates with λ equal to three, which should yield approximately a 50% advantage versus low-strength steels. However, new architectures along with AHSS for the door outer should allow for a 15 to 20% reduction while keeping the door in steel. The difference between aluminum and steel for the same architecture and using AHSS should be between 25 and 35%. The hybrid gave the same 48% improvement as the aluminum intensive design. In light of the next generation fuel economy and safety regulations, it appears that aluminum may grow its applications for closures. In fact, a large percentage of hoods and some decklids are using aluminum for high-volume vehicle applications today. However, steel could improve its position on doors if a better way of integrating the door with the rest of the body structure could be found. There are a couple of examples in the marketplace where the door has been designed to help with side impact or roof crush, but there are no designs that can serve to significantly improve the overall vehicle stiffness beyond what is available with the conventional hinge and latch systems.

		Baseline Vehicle Mass (kg)	Front Doors Frame					Mass Saving for 2 doors (kg)	Incremental from Option 1 $/kg
			Mass Reduction			Cost Increase			
			%	LWV Mass (kg)	Mass Saving (kg)	Incremental ($)	Mass Saving Premium ($/kg)		
Option 1	AHSS	16.4	-15%	13.9	2.5				
Option 2 Eng Sol	Aluminum Stampings	16.4	-48%	8.45	7.95				
Option 3	Aluminum Outer Stamping	5.6	-52%	2.7	2.9				
	Magnesium Casting Inner	6.5	-49%	3.3	3.2				
	Other Parts (Aluminum)	4.3	-40%	2.6	1.7				
	Total	16.4	-48%	8.6	7.8				

Figure 4.19 Front doors analysis

In conclusion, the AHSS intensive body with aluminum closures and hot-stamped steel bumpers provided the high-value alternative for the NHTSA LWV study. The baseline 435.9 kg for these subsystems would be reduced to 309.2 kg for a 29.1% weight reduction. A distribution of materials for the body structure and closures is shown in Figure 4.20, and a more detailed look at some of the significant high-strength steel parts of the body structure is shown in Figure 4.21. Note the extensive use of hot stamping in Figure 4.21. This reinforces the view that hot stamping is and will continue to be the fastest growth area for AHSS enablers.

The LWV chassis (Figure 4.22) came out to be 28.6% lighter than the baseline. At nearly the same weight the redesigned AHSS suspension arms (both front and rear) came out to be about the same weight as the redesigned aluminum arms, but at much less cost. Therefore, the EDAG team went with AHSS. This was consistent with a 2010 A/SP study. The redesigned front subframe came out to be only slightly lighter in aluminum versus AHSS, but at a much higher cost. In this case, EDAG went with the aluminum design. Similarly, the team went with aluminum for the rear subframe. However, for both the front and rear subframes, AHSS was very competitive from a weight reduction perspective and at much

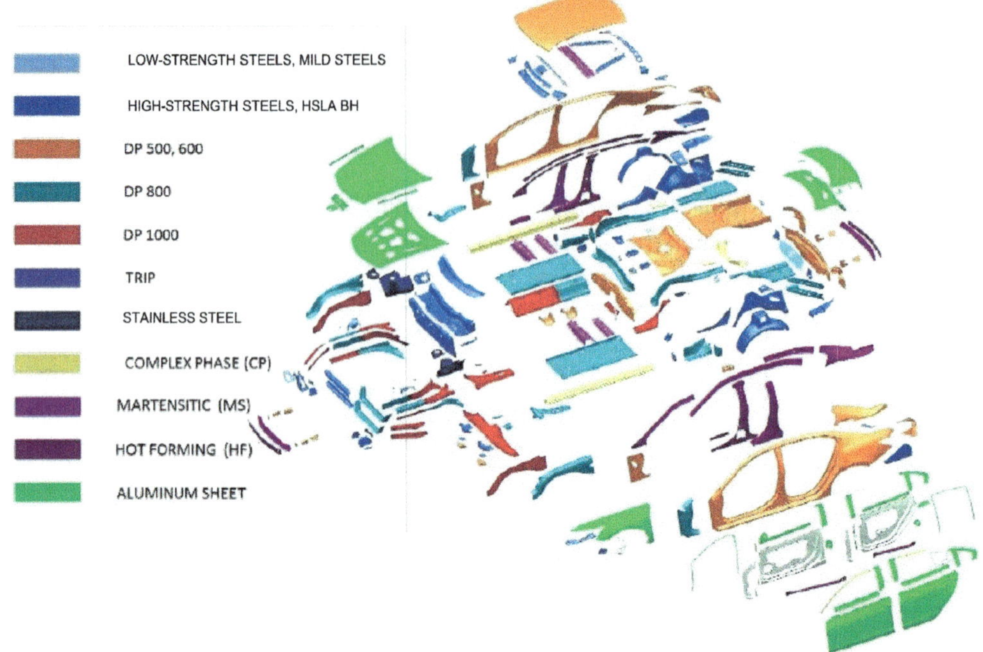

Figure 4.20 Body and closures material selection

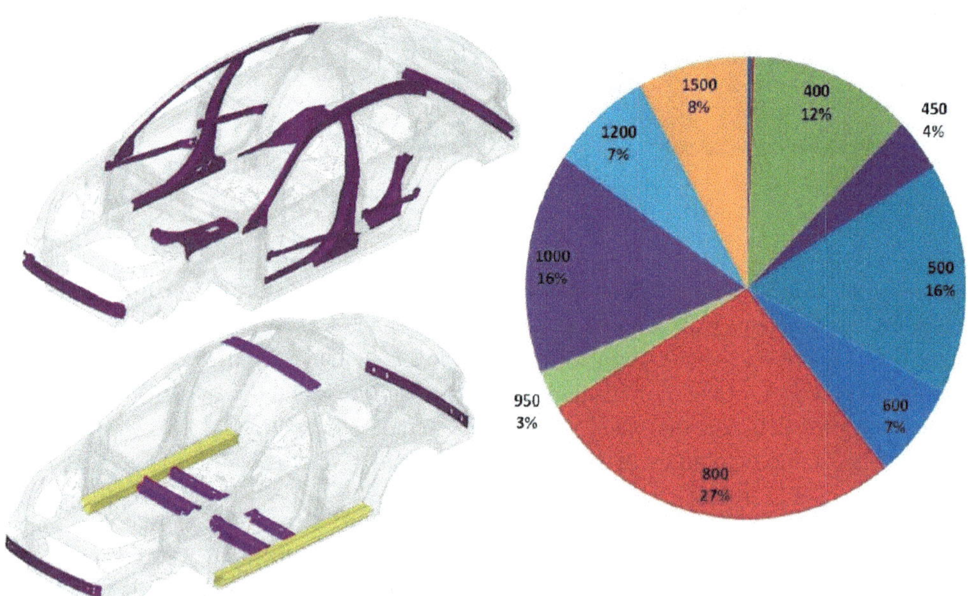

Figure 4.21 Body-in-white AHSS grades in terms of tensile strength

lower cost. Conversion of the four wheels from conventional high-strength steel (CHSS) to AHSS resulted in a 15% mass saving, and the LWV wheels stayed in steel. Though the final NHTSA report on the LWV study did not go into detail on the aluminum versus steel comparison for the wheels, there have been a number of good steel versus aluminum wheel presentations given in recent years [4-6]. The difficulty for competitive wheel weights in steel has been that the cast aluminum wheels can have a very stylish surface versus having to use a wheel cover with the steel wheel. However, the referenced study shows that with AHSS the weight of the steel wheel can be reduced to be less than the aluminum design to accommodate the addition of a stylish steel wheel cover. At equivalent net weight to aluminum the steel design comes in at about 30% lower cost.

Figure 4.22 Light weight vehicle chassis

One of the outstanding characteristics of the LWV EDAG study was the amount of topology optimization that was used. The procedure started with an exterior and interior scan of the baseline vehicle (Figure 4.23).

After scanning, a topology optimization solid model is created (Figure 4.24).

This model was analyzed using an optimization program called OPTISTRUCT, which was developed by ALTAIR Engineering, Inc. The following load cases were used for this analysis:

1. Stiffness Bending and Torsion
2. Frontal NCAP Full Barrier
3. IIHS 40% ODB Front Crash
4. IIHS Side
5. FMVSS No. 214 (Pole IMPACT)
6. FMVSS No. 301 (Rear Crash)
7. FMVSS No. 216 (Roof Crush

Figure 4.23 Topology solid volume generation

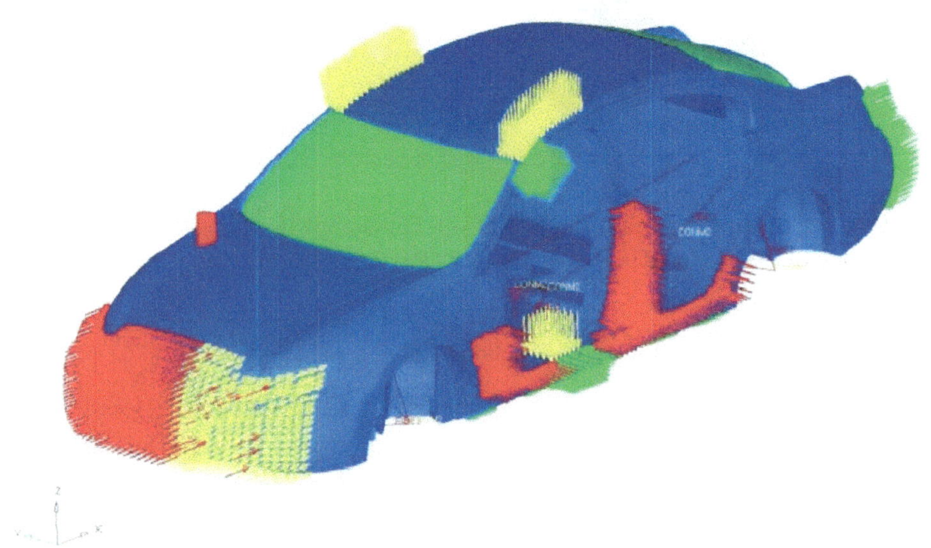

Figure 4.24 Topology optimization model

Figure 4.25 shows the load paths determined by OPTISTRUCT. The pink load paths were those directly determined by OPTISTRUCT. The gray load paths were chosen from the pink load paths as being design feasible. For instance, the pink load paths on the roof would have prevented a sun roof from being installed. Therefore, a production feasible set of gray load paths was utilized for further refinement.

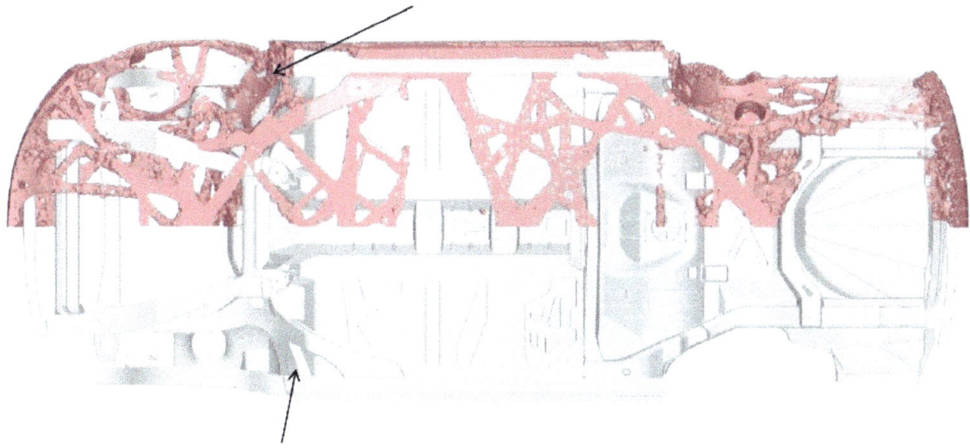

Figure 4.25 Topology optimization results

To take advantage of the topology optimization, decisions had to be made on the acceptable material, steel processing, and technology portfolio available to the design team. The material portfolio for steel was shown in Figure 4.17. The processing portfolio included the following:

1. Conventional coils
2. Laser-welded coils
3. Tailor rolled coils
4. Conventional blanks
5. Laser-welded blanks
6. Tailor rolled blanks
7. Laser-welded tubes
8. Variable walled tubes
9. High-frequency induction welded tubes
10. Variable walled profiles

The technology portfolio considered was very similar to our list of enabling technologies, which was discussed in the last chapter. The next step (termed 3G optimization) for the computer optimization process was to develop optimum grades, thicknesses, and cross sections for the members, which were coincident with the gray load paths. Three computer programs were set up to run in a continuous loop to accomplish this task:

1. HEEDS, an optimization program, administered by Red Cedar Technologies
2. SFE Concept, a program that would translate the variables determined by HEEDS into a finite element model
3. LSDYNA (LSTC, Inc.) a nonlinear finite-element solver

Figure 4.26 shows an overview of the steps in the 3G optimization process, along with an optimized rocker section.

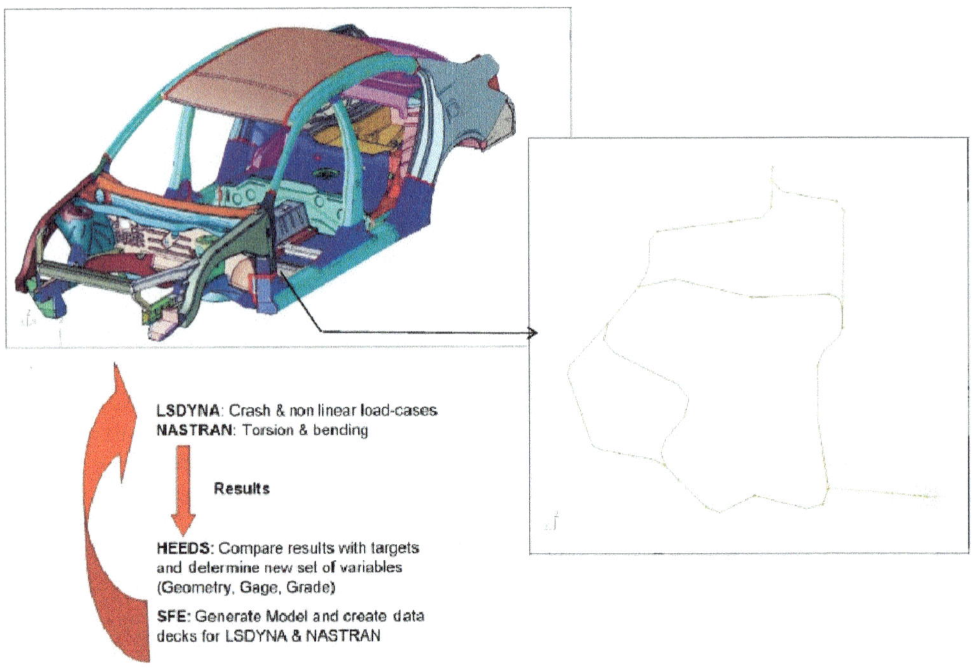

Figure 4.26 3G optimization process

Figure 4.27 shows a comparison of the baseline Accord rocker section, in red, and the optimized rocker in gray.

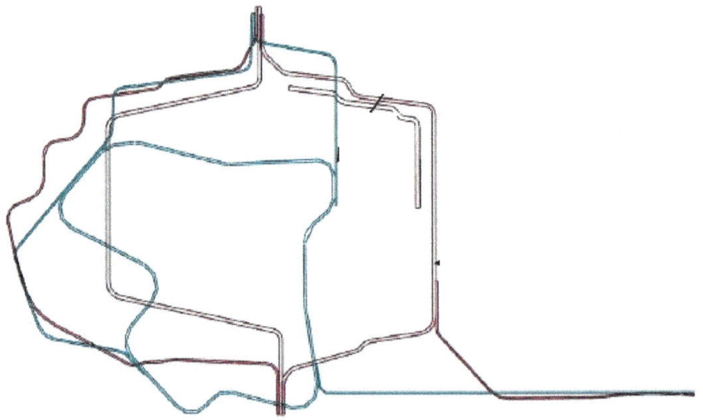

Figure 4.27 3G optimized-section comparison

After the optimization process was complete, a comparison was made of the LWV to the baseline Accord for the following critical load cases:

1. Torsional and Bending Stiffness
2. Normal Mode Frequencies
3. USNCAP Frontal Rigid Barrier 35 MPH test
4. IIHS Offset Barrier 40 MPH Deformable Barrier Test
5. USSINCAP Lateral Side Impact Test
6. IIHS Side Impact 50 km/h Test
7. NCAP Rigid Side Pole 20MPH Test
8. IIHS Roof Crush Test
9. Rear 301 Fuel Tank Integrity 50 MPH Test

Figure 4.28 shows the linear NASTRAN results for the torsional stiffness, which is a critical static load case. The optimized LWV is shown to be significantly higher than the baseline Accord, which is highly desirable.

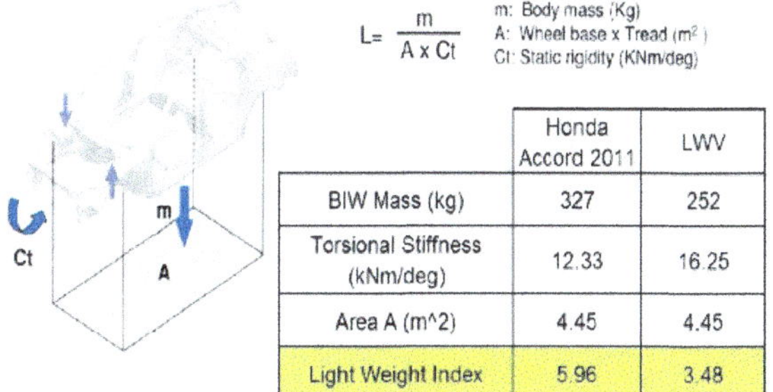

	Honda Accord 2011	LWV
BIW Mass (kg)	327	252
Torsional Stiffness (kNm/deg)	12.33	16.25
Area A (m^2)	4.45	4.45
Light Weight Index	5.96	3.48

Figure 4.28 LWV versus Accord for torsional stiffness

Figure 4.29 shows a comparison of the Accord test vehicle and the LWV computer model about to strike the rigid barrier for the 35 MPH frontal impact event. The LS-Dyna computer results versus the Accord test results are shown in Figure 4.30. The decelerations are essentially the same, which would indicate that the dummy loads (i.e., head index criteria and chest Gs) would be about the same. Also, the deformation at the dash and brake pedal is less for the LWV as shown at the bottom-right chart, which is desirable.

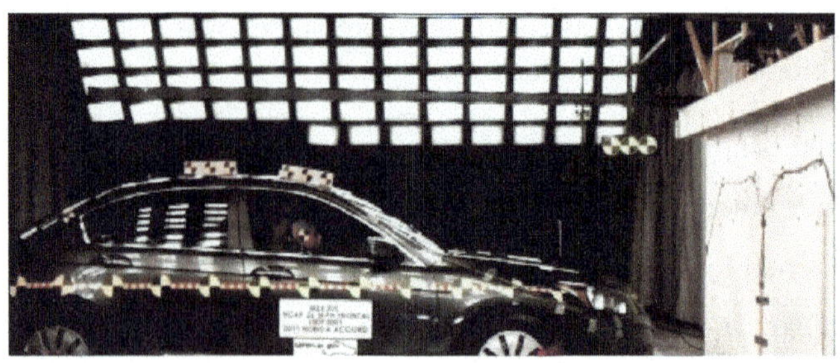

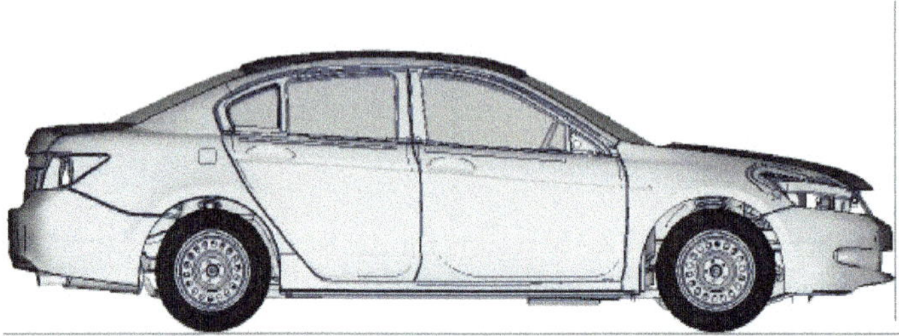

Figure 4.29 35 Accord test vehicle and LWV computer model

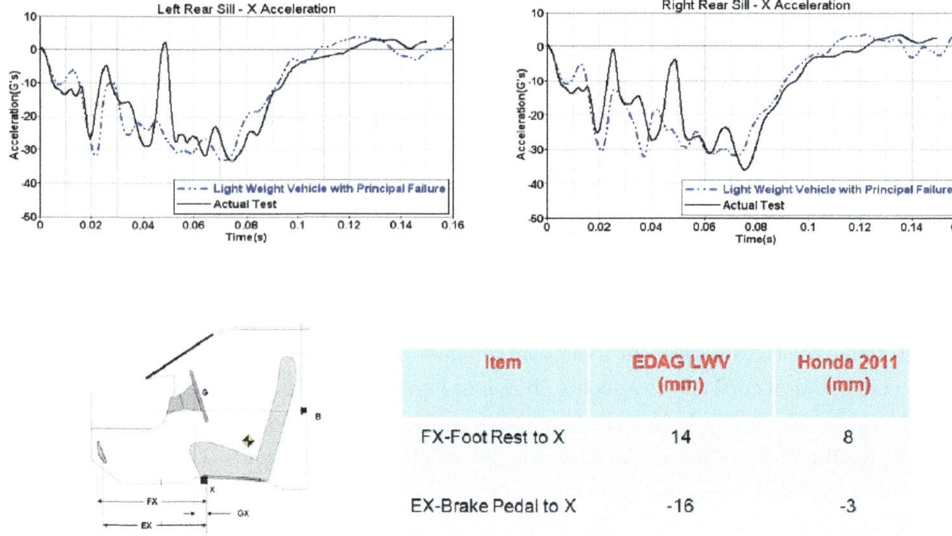

Item	EDAG LWV (mm)	Honda 2011 (mm)
FX-Foot Rest to X	14	8
EX-Brake Pedal to X	-16	-3

Figure 4.30 Accord test results versus LWV computer results for 35 MPH full frontal impact

Figure 4.31 shows a comparison of Accord test results versus LWV computer results for the IIHS deformable offset barrier test. As shown in the upper-left graph, the decelerations are essentially the same; and as shown in the lower right, the intrusions in the passenger compartment are essentially the same.

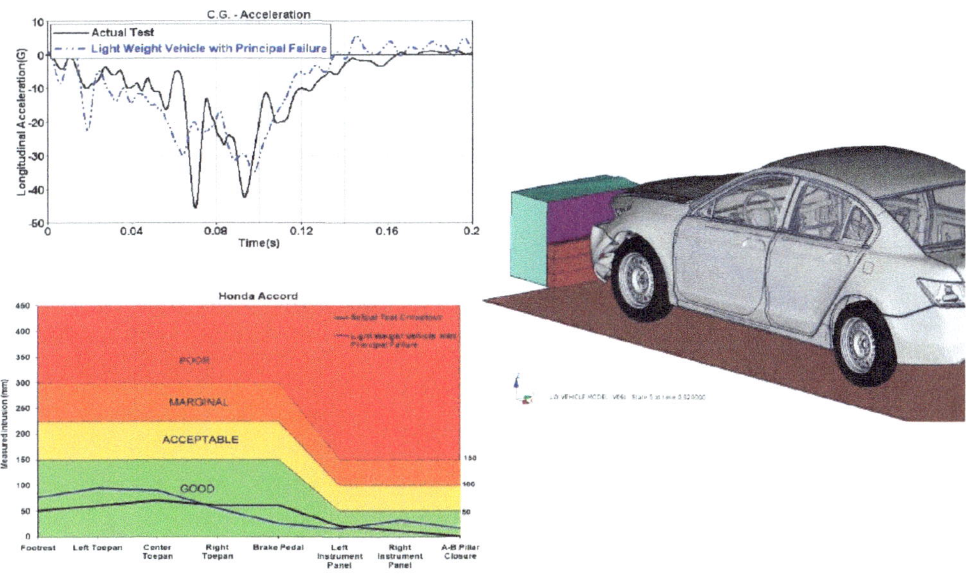

Figure 4.31 IIHS 40MPH deformable offset barrier comparison

Figure 4.32 shows the pictures of the LWV computer model (left side) and the physical Accord after the IIHS side impact 50 KPH event. Visually the computer model and the Accord look about the same. Figure 4.33 shows the numerical results. The accelerations and the intrusions look to be about the same on the left-hand side of Figure 4.33, and all the intrusions for both the Accord test results and the LWV results are within the green zone in this figure.

Figure 4.32 IIHS side impact 50 KPH test

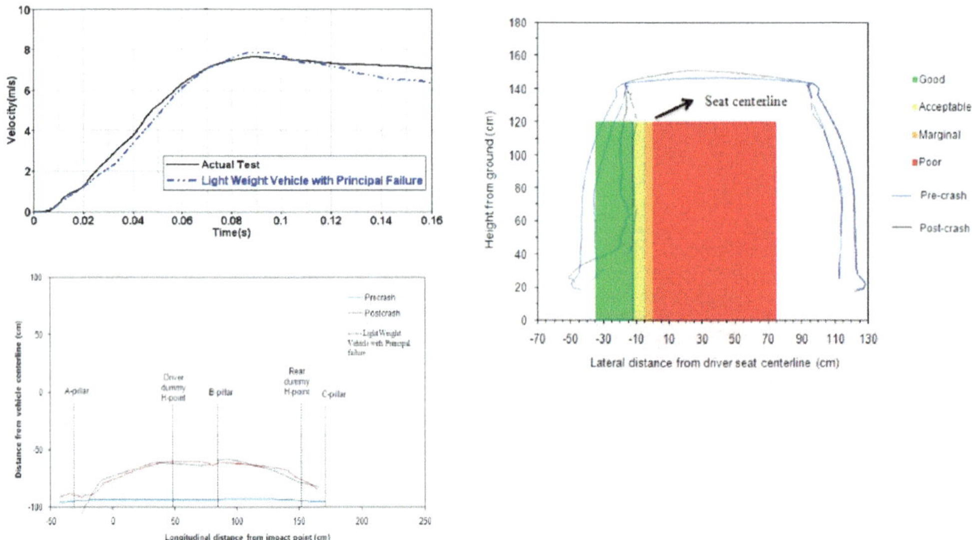

Figure 4.33 Numerical results for IIHS side impact 50 KMH test

Figure 4.34 shows the LWV computer results versus the Accord test for the NCAP Rigid Side Pole 20 MPH test. Again, the comparison is about the same.

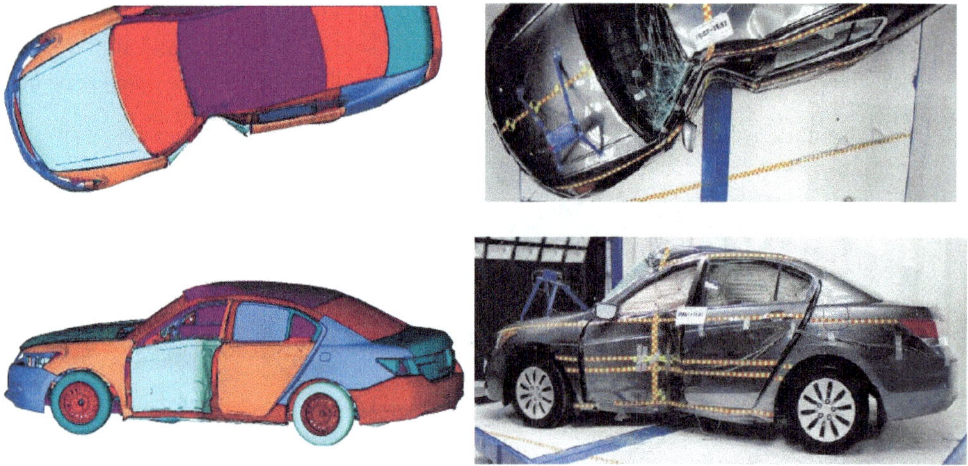

Figure 4.34 NCAP rigid side pole 20MPH

Figure 4.35 shows the comparison of the LWV computer simulation to the physical test of the Accord for the IIHS roof crush test. The graph at the left-bottom shows the comparison of the two for the platen force, which is applied diagonally to the roof, as a function of the platen displacement. In the curves on the right, the original curves are replotted as SWR (strength to weight ratio) as a function of platen displacement. SWR is the force divided by the weight of the vehicle. The curves look somewhat different in that the LWV weighs less and therefore the curve is higher. The IIHS requirement is that SWR must reach 4 somewhere before 5 inches of displacement. For this load case, the LWV outperformed the Accord (green versus yellow).

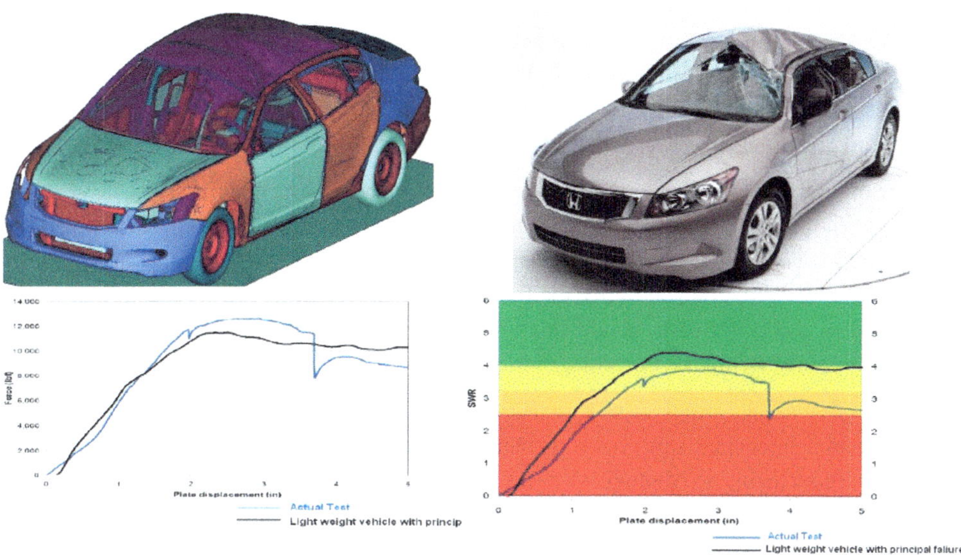

Figure 4.35 IIHS roof crush test

Finally Figure 4.36 shows the simulation of the LWV for the fuel tank integrity test. The LWV passed this test also.

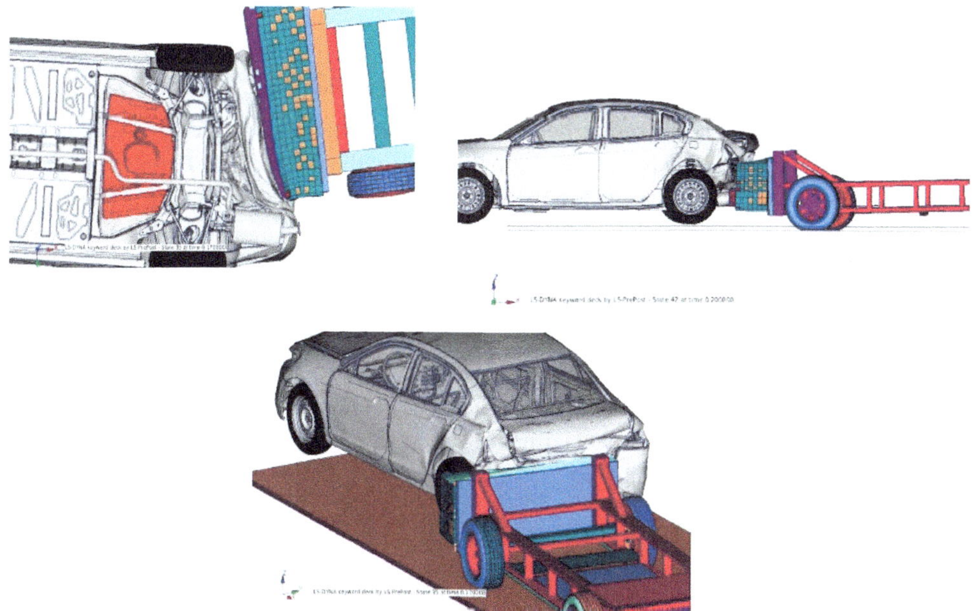

Figure 4.36 Rear 301 fuel integrity 50 MPH test

The following were overall conclusions of this study:

1. The study demonstrated that mass reductions of up to 23% are likely feasible while still maintaining performance and safety functionality and costs at plus or minus 10% of the base vehicle. Most notable as far as steel is concerned, the body-in-white was kept in steel at a 23% weight reduction, and much of the chassis was retained in steel. The subframes, though converted to aluminum for the LWV, could have been done in AHSS without an appreciable decrease in the overall vehicle weight reduction.
2. Bear in mind that this study was done for a four-door mid-size sedan. The relative weight reductions of steel and aluminum body-in-white could change for different types of vehicles. For instance, for a full-sized pickup, one might suspect that aluminum could have a greater advantage because the frame carries more of the noise, vibration, and harshness and crash loads. However, fatigue plays a bigger role for this type of vehicle, and that could swing the percentages in favor of steel in that it has a defined endurance limit.
3. The approach for the study was an evolutionary implementation of advanced materials and manufacturing technologies currently used in the automotive industry.
4. The recommended materials (AHSS, aluminum, magnesium, and plastics), manufacturing processes (stamping, hot stamping, die casting, extrusions, roll forming, and hydroforming), and assembly methods (spot welding, laser welding, and adhesive bonding) are used today, some to a lesser degree than others.
5. The recommended technologies should be able to be fully developed within the normal product development design cycle using the current design and development methods prevalent in the automotive industry.

4.5 Relative Costing of AHSS and Alternative Materials

The report referenced in the preceding section [4-5] has the costing data included in chapter 9 of that report. The data, which is pertinent to the topic of this book, will be summarized here. All of the costing data presented in the referenced report is in 2010 USD. The sources of the OEM manufacturing costs were technical cost modeling (TCM) and supplier assessment. The TCM approach was developed at Massachusetts Institute of Technology by the Materials Systems Laboratory researchers [4-7]. TCM was initially developed to support the World Auto Steel ULSAB-AVC Project. It has also been employed by the U.S. Department of Energy for costing exercises for its Vehicle Technologies Program. Only direct costs for manufacturing the parts and the assembly of the parts were considered in the NHTSA/EDAG LWV program. The major cost elements directly linked to manufacturing and assembly are summarized as follows [4-5, p. 361]:

1. Fabrication costs of all the parts including tooling costs
2. Assembly costs, including tooling costs
3. Material
4. Direct labor

5. Energy
6. Equipment
7. Building (for manufacturing and assembly
8. Maintenance (for manufacturing and assembly)
9. Overhead labor in manufacturing plant (i.e., indirect labor directly connected to the manufacturing and assembly process)

For this study, the cost model was created based on the assumption that the parts are manufactured in a Greenfield facility (or a facility new from the ground up) in the United States. The cost assessment encompasses the raw material (steel, aluminum alloy, and so on) entering the plant to the complete vehicles leaving.

The LWS material prices were based on the average of North American 2011 material prices (www.platts.com), which were converted to 2011 prices using the gross domestic product deflator. For materials, which were not available through published sources, industry experts were used. For mild, cold rolled steel, the published price was $0.93 USD per kilogram. In Figure 4.37, the reference steel price is shown along with the premium for other grades of steel used in the NHTSA/EDAG study. This table came from the previously referenced report. The premiums are also shown for hot-dipped galvanized (HDG) coated, exposed, Tailor rolled coils, tubes, and multiwall tube blanks. World Auto Steel supplied the premiums for the higher grades of steel. Similarly, the aluminum prices were determined. Some significant information can be gleaned from the table by looking at the following entries:

1. Mild, cold rolled reference: $0.93/kg or $0.42/lb
2. HSLA 350/450: $1.05/kg or $0.48/lb
3. DP 350/600: $1.19/kg or $0.54/lb
4. HF 1050/1500 (aluminized): $1.65/kg or $0.75/lb
5. Austenitic stainless steel: $4.65/kg or $2.10/lb
6. Average aluminum sheet costs: $4.71/kg or $2.14/lb

The CHSS (HSLA 350/450) has about 13% higher price than mild steel, and the first step up in grade from HSLA 350/450 to AHSS DP 350/600 incurs about a 13% increase over the HSLA. The hot-formed (HF), aluminized steel is a big jump up in price from the dual phase, but the non-aluminized HF is more in the range of the dual-phase (DP) steels. Austenitic stainless is very high in price, which accounts for the fact that it hasn't made big inroads into the automotive structural market. Austenitic stainless can be considered second generation AHSS and has very high strength and elongation. The problem with austenitic stainless is that it is almost the cost of aluminum. Another interesting fact is that aluminum is over 4.5 times the cost of a mix of mild and CHSS. This is related to the fact that it takes 4.5 times the energy to produce a pound of aluminum versus a pound of steel.

Item #	Steel Grade	Min t (mm)	Max t (mm)	Ref Material Price ($/kg)	Grade	HDG	Visible	Tailor Rolled Coil	Tubes (straight, as shipped)	Multiwall Tube Blank
					Premium ($/kg)	Premium ($/kg)	Premium ($/kg)	Premium ($/kg)	Premium ($/kg)	Premium ($/kg)
	Reference US Spot Midwest Market Price Trend 2009									
1	Cold Rolled Reference - Mild 140/270	0.35	4.60	0.93	0.00	0.06	0.05	0.55	0.25	0.65
2	BH 210/340	0.45	3.40		0.05	0.06	0.10	0.55	0.25	0.65
3	BH 260/370	0.45	2.80		0.05	0.06	0.10	0.55	0.25	0.65
4	BH 280/400	0.45	2.80		0.07	0.06	0.10	0.55	0.30	1.10
5	IF 260/410	0.40	2.30		0.07	0.00	0.10	0.55	0.30	0.70
6	IF 300/420	0.50	2.50		0.10	0.00	0.10	0.55	0.30	1.10
7	HSLA 350/450	0.50	5.00		0.12	0.10	NA	0.55	0.30	1.50
8	HSLA 420/500	0.60	5.00		0.14	0.10	NA	0.55	0.45	1.25
9	HSLA 490/600	0.60	5.00		0.16	0.10	NA	0.55	0.45	1.65
10	HSLA 550/650	0.60	5.00		0.35	0.10	NA	0.55	0.45	1.65
11	HSLA 700/780	2.00	5.00		-	-	-	-	-	-
12	SF 570/640	2.90	5.00		0.35	0.10	NA	NA	0.45	2.05
13	SF 600/780	2.00	5.00		0.35	0.10	NA	NA	0.45	2.05
14	TRIP 350/600	0.60	4.00		0.40	0.10	NA	NA	0.45	1.25
15	TRIP 400/700	0.60	4.00		0.45	0.10	NA	NA	0.45	1.65
16	TRIP 450/800	0.60	2.20		0.50	0.10	NA	NA	0.50	1.30
17	TRIP 600/980	0.90	2.00		0.55	0.10	NA	NA	0.55	1.35
18	FB 330/450	1.60	5.00		0.20	0.10	NA	0.55	0.30	1.10
19	FB 450/600	1.40	6.00		0.25	0.10	NA	0.55	0.45	1.65
20	DP 300/500	0.50	2.50		0.20	0.10	0.10	0.55	0.45	0.85
21	DP 350/600	0.60	5.00		0.26	0.10	0.10	0.55	0.45	1.25
22	DP 500/800	0.60	4.00		0.31	0.10	NA	0.55	0.50	0.90
23	DP 700/1000	0.60	2.30		0.38	0.10	NA	NA	0.55	0.95
24	DP 800/1180	1.00	2.00		-	-	-	-	-	-
25	DP 1150/1270	0.60	2.00		0.38	0.10	NA	NA	0.55	0.95
26	CP 500/800	0.80	4.00		0.31	0.10	NA	NA	0.50	1.30
27	CP 600/900	1.00	4.00		0.35	0.10	NA	NA	0.52	1.32
28	CP 750/900	1.60	4.00		0.40	0.10	NA	NA	0.52	1.32
29	CP 800/1000	0.80	3.00		0.45	0.10	NA	NA	0.55	1.35
30	CP 1000/1200	0.80	2.30		0.47	0.10	NA	NA	0.60	1.40
31	CP 1050/1470	1.00	2.00		0.47	0.10	NA	NA	0.60	1.80
32	MS 950/1200	0.50	3.20		0.47	NA	NA	NA	0.60	1.00
33	MS 1150/1400	0.50	2.00		0.48	NA	NA	NA	0.60	1.40
34	TWIP 500/980	0.80	2.00		1.20	0.10	NA	NA	0.60	1.80
35	MS 1250/1500	0.50	2.00		0.51	0.10	NA	NA	0.65	1.05
36	HF 1050/1500 (22MnB5)	0.60	4.50		0.75	NA	NA	0.55	0.65	1.05
37	Al 6111 Exposed			4.97	4.04					
38	Al 6111			4.90	3.97					
39	Al 6061			5.13	4.20					
40	Al 5182			4.83	3.90					
41	[illegible]				[illegible]	0.10	NA	NA	NA	NA
42	Stainless Seel				3.72	0.10	NA	NA	NA	

Figure 4.37 Material costs for the NHTSA/EDAG LWV study

From this table and some of the other data in the report, you could conclude that to produce the conventionally stamped parts on the body the material cost would be about $1.00/kg, and the extra manufacturing cost would be about $2.60/kg for a total part cost of $3.60/kg. For the hot-stamped parts, the material cost would be $1.65/kg and the extra manufacturing cost would be about $5.00/kg for a total hot-stamped parts cost of $6.65/kg. Since the processing costs are so much larger for the hot-stamped parts versus the conventionally stamped parts, a virtual material cost for the hot stamped material can be derived by assuming that the processing costs are the same for the stamped versus the hot-stamped parts by subtracting $2.60/kg from $6.65/kg. Then the virtual material cost for the hot-stamped material is $4.05/kg or $1.84/lb. This may be extrapolating the data from the NHTSA/EDAG study a bit too far, in that there was only 10 kg of hot-stamped parts in the body-in-white. However, it does show that the relative costs of hot-stamped material to conventional steels is considerably higher than Figure 4.37 would indicate if taking into account the manufacturing costs.

In terms of the manufacturing cost adding to the material costs, EDAG developed the chart in Figure 4.38 from the data developed in the LWV study. For all the materials considered in the NHTSA/EDAG LWV study, the costs of the material plus the manufacturing costs were significantly higher than the material costs alone. Since the first three rows are of primary interest for this book, it is interesting to note that the material plus manufacturing costs for the mild steel and CHSSs are 28% higher than for the material cost alone. For AHSS it is 44%, and for aluminum sheet it is 32%. The increase in the cost of die materials and in die operations accounts for AHSS high manufacturing costs as a percentage of material costs. However, the net aluminum costs are still 2.7 times the AHSS costs.

Material	Material Cost $/kg	Reference	Material Cost with Manufacturing $/kg	Manufacturing Process Scrap	Manufacturing Difficulty	Scrap Return Premium $/kg
Steel upto 590 MPa strength - Average	1.14	Platts - WorldAutoS	1.46	0.45	1.00	0.44
Steel AHSS Average	1.44	Platts - WorldAutoS	2.08	0.45	1.10	0.44
Aluminum Sheet	4.26	Platts	5.62	0.45	1.10	2.38
Aluminum Cast	2.54	Platts	3.31	0.03	1.30	2.22
Magnesium Cast	4.98	Platts	6.57	0.03	1.30	2.44
Vynel Ester Compound	4.24	DAS 2010	5.13	0.10	1.10	0.00
Fiber Glass	1.50	DAS 2010	2.70	0.20	1.50	0.00
Carbon Fiber	17.60	DAS 2010	42.24	0.20	2.00	0.00
Gray Iron/steel	1.50	Supplier	2.02	0.05	1.30	0.44
SMC	3.00	Supplier	4.10	0.05	1.30	0.00

Figure 4.38 LWV material costs normalized for manufacturing

At the 2013 GDIS Conference, Jody Shaw of the U.S. Steel Corp. gave a talk [4-8], which provided some good summary points for EDAG's 2012 presentation of the NHTSA/EDAG LWV study and extended the data to look at how the materials choices would impact emissions. His slide that provided a good overview of the weight choices and selections for the LWV is shown in Figure 3.39. As can be seen in the chart, there were three options (i.e., all AHSS, all aluminum, or a combination of aluminum, magnesium, and carbon fiber reinforced polymer) to choose from for each system of the vehicle. Remember that since the LWV was targeted for model year 2020, the last column was not judged to be a viable alternative, as the carbon fiber alternative was judged to have cost and availability issues in that time frame. The alternative that was picked for the body structure was AHSS based on 22% weight reduction and low cost. For the other systems, aluminum was chosen for the closures, AHSS for the bumpers, aluminum for the suspension system, AHSS for the wheels, plastics for the seats, and magnesium for the cross car IP beam. The LWV achieved a 22% weight reduction for these selections. It is interesting that the total vehicle, minus the powertrain, only achieved a 3% improvement over the AHSS vehicle.

Another interesting summary slide from Mr. Shaw's presentation is shown in Figure 4.40. Here the various material alternatives are lumped in terms of structural subsystems, powertrain, and nonstructural subsystems. This chart shows that the only real mass differences between the vehicle choices were for the structural components.

Material Selection of the LWV Program

System / Mass(Kg) / % Saved	Baseline (Kg)	All AHSS (Kg)	All AHSS (%)	Light Weight Vehicle (Kg)	Light Weight Vehicle (%)	All Aluminum (Kg)	All Aluminum (%)	Aluminum Magnesium CFRP (Kg)	Aluminum Magnesium CFRP (%)
Non-Structural	370	325	12%	325	12%	325	12%	325	12%
Powertrain	347	281	19%	281	19%	281	19%	281	19%
Body Structure	328	255	22%	255	22%	213	35%	164	50%
Closures	92	78	15%	48	48%	48	48%	48	48%
Bumpers	16	9	45%	9	45%	9	45%	9	45%
Suspension	140	91	35%	87	38%	87	38%	87	38%
Wheels	94	79	15%	79	15%	79	15%	79	15%
Seats	67	60	10%	47	30%	47	30%	49	27%
IP Beam	32	22	30%	22	30%	22	30%	22	30%
Total	1480	1197	19%	1149	22%	1108	25%	1060	28%

Steel
AHSS
Aluminum
Magnesium
Plastics
CFRP

Figure 4.39 Material selection of the LWV program

Material Content of LWV Program

Mass Differences in Structural Materials Only

Mass (Kg)
1600
1400
1200
1000
800
600
400
200
0
Base-line
AHSS Intensive
LWV AHSS BIW Alum Closures Alum Chassis
Alum. Intensive
Alum. Mag. CFRP
Structural: CFRP, Cast Magnesium, Stuct. Cast Alum, Alum Sheet, AHSS, Convensional Steel
Powertrain: Cast Alum, Cast Iron/Steel Bar
Non-structural: Glass, Copper, Plastics, Fluids, Misc

Figure 4.40 Material content of the LWV program

In Figure 4.41, Mr. Shaw showed a fuel economy comparison for the four vehicle choices versus the baseline 2011 Honda Accord. It is interesting to note that the LWV gained less than a half of a mile per gallon over the AHSS intensive vehicle. As stated previously, it takes a lot of weight reduction to provide significant fuel economy benefits, even with a resized powertrain.

U. S. Steel **Fuel Economy of LWV Program**

Fuel Economy Improvements Result from Mass Reduction and Resized Powertrain (also enabled by mass reduction)

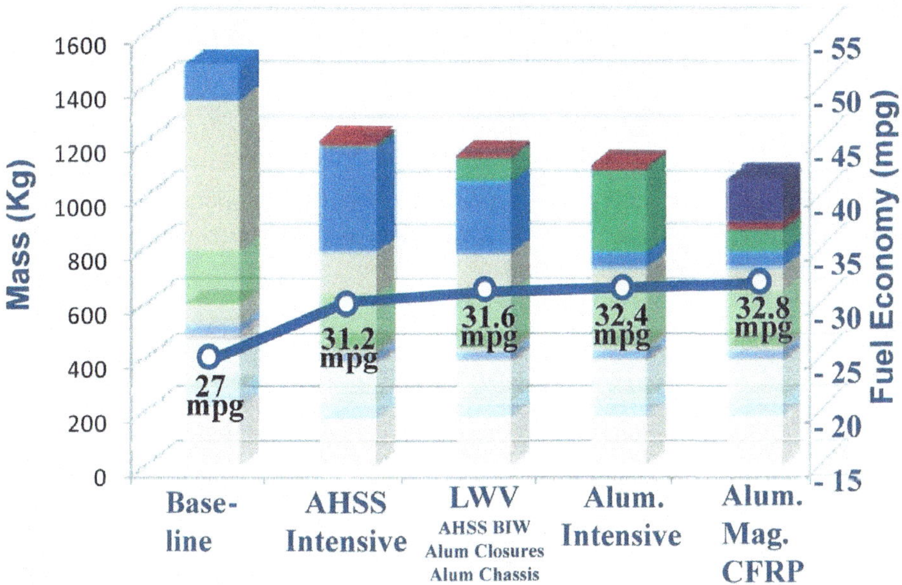

Figure 4.41 Fuel economy of the LWV program

In Figure 4.42, Mr. Shaw showed the relative manufacturing cost premium for the four alternatives, and on that basis shows the advantages/disadvantages to the customer of the three other alternatives relative to the AHSS intensive vehicle. Significant is the fact that the AHSS intensive vehicle is $210 less than the LWV and $820 less than the aluminum intensive vehicle. If the increase in cost is passed on to the customer, it will take the LWV customer 8.6 years to break even over the AHSS intensive customer, and it will take the aluminum intensive vehicle customer 11.5 years to break even over the AHSS intensive vehicle customer.

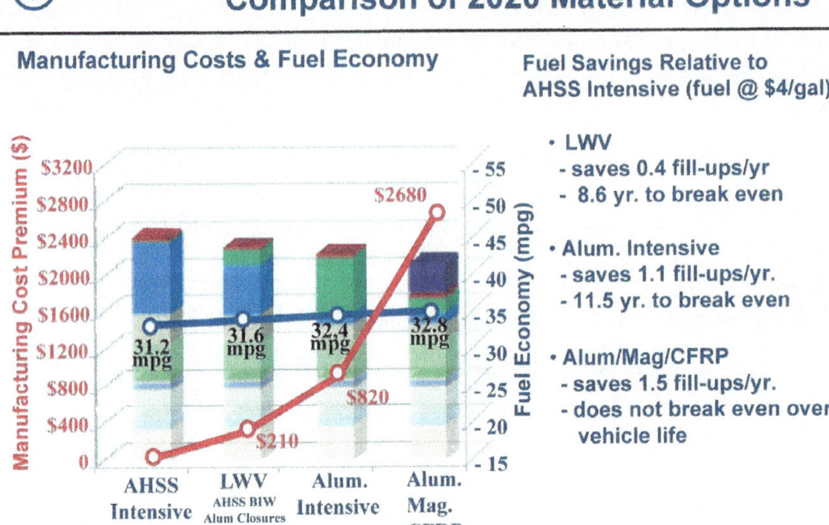

Figure 4.42 Comparison of 2020 material options

Jody Shaw then went on to do a life cycle assessment for the different options. For an accurate life cycle assessment, it is important to take into account all the sources of CO_2. The contributing factors are shown in Figure 4.43.

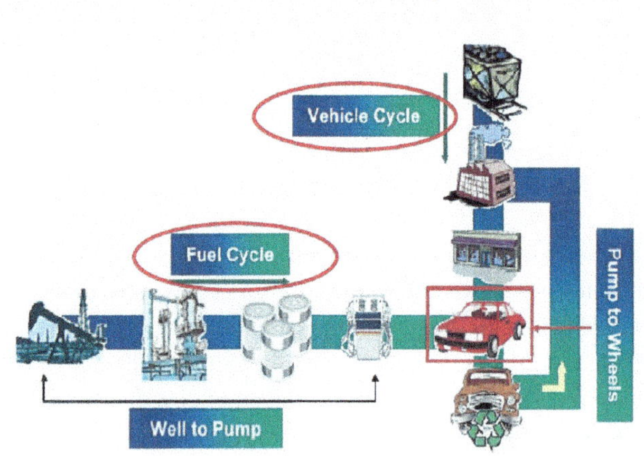

Figure 4.43 Life cycle assessment of CO_2 reduction

The first factor to consider is the material production. A comparison in terms of CO_2 per kg of material for various automotive materials is shown in Figure 4.44. Note that steel is much lower than the other alternative materials.

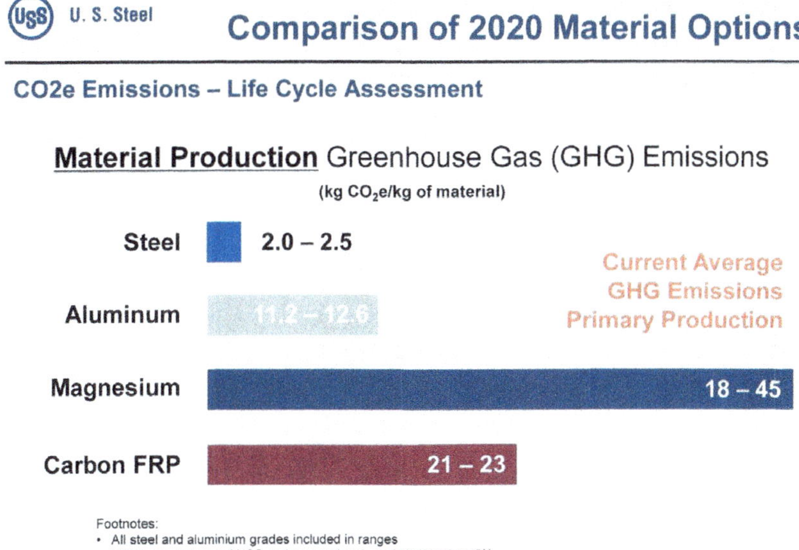

Figure 4.44 Greenhouse gas emissions for automotive material production

In Figure 4.45, Mr. Shaw shows the emissions for the four different alternatives of the LWV study during the manufacturing stage. The greenhouse gas emissions are over three times the AHSS intensive vehicle to manufacture the aluminum intensive vehicle and the aluminum/magnesium/CFRP vehicle.

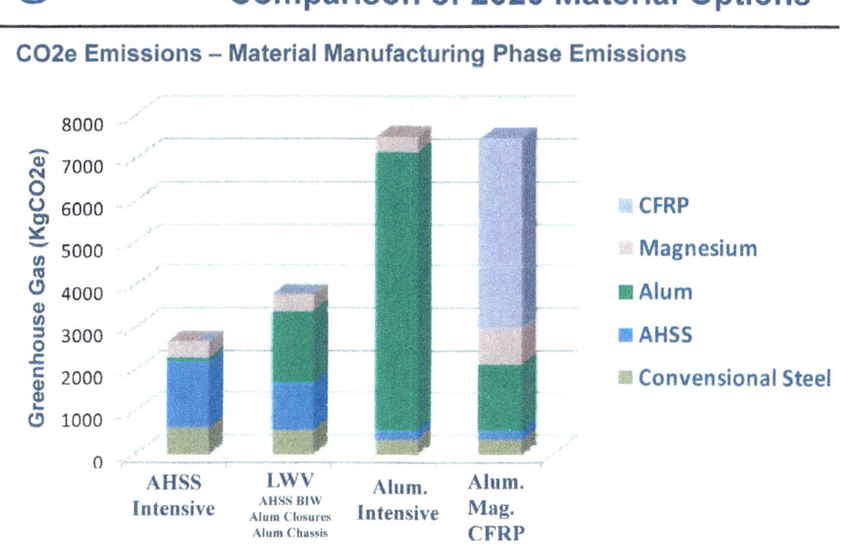

Figure 4.45 Manufacturing phase emissions

In Figure 4.46, the total greenhouse gas emissions are shown for manufacturing, driving, and end-of-life. It can be seen from this chart that the AHSS intensive vehicle generates the lowest greenhouse gas emissions throughout the useful life of the vehicle. The differences between the four alternative vehicles appear much more pronounced if plotting the lighter three alternatives relative to the AHSS intensive vehicle's greenhouse gas emissions.

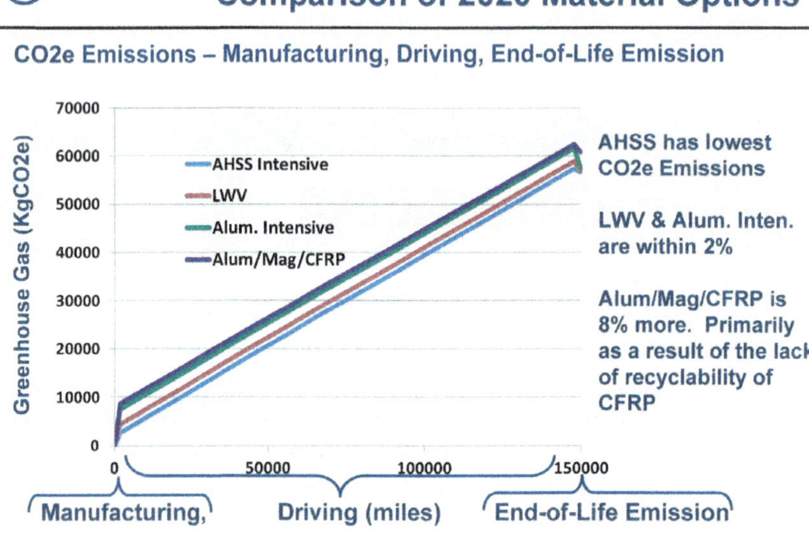

Figure 4.46 Manufacturing, driving, and end-of-life emissions

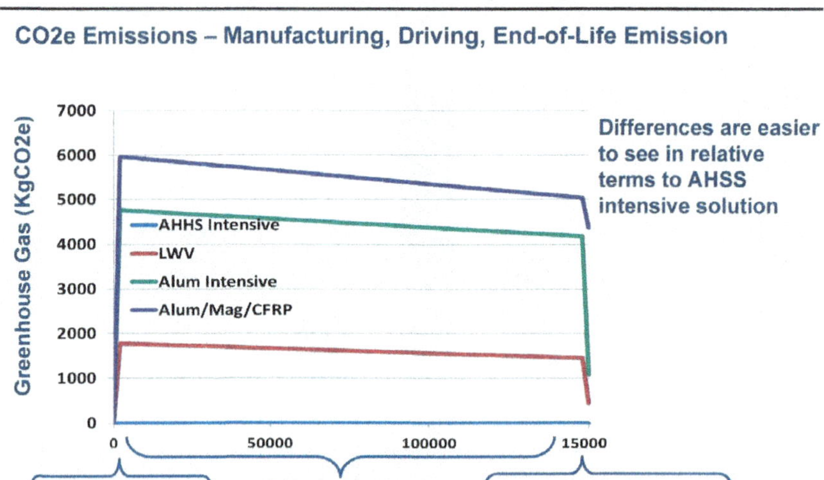

Figure 4.47 Total greenhouse gas emissions relative to the AHSS intensive vehicle

In summary, Mr. Shaw's presentation makes a strong case for the AHSS intensive alternative, in that the LWV only gains .4 MPG, will take 8.6 years to recover its price premium, and significantly increases the CO_2 emissions. However, the goal of the NHTSA/EDAG study was to develop a technically feasible vehicle, which moves the industry in a direction of the 2025 U.S. fuel economy mandate. From this perspective, the LWV might be the best alternative in that it does deliver better fuel economy than the AHSS intensive vehicle at slightly more cost. From any point of view, the aluminum intensive vehicle is much more costly and the aluminum/magnesium/CFRP vehicle cannot make it by 2025.

4.6 References

4-1. Popov, E. P, 1964, *Mechanics of Materials*: Prentice-Hall, Inc., Englewood Cliffs, NJ.

4-2. Harris, Hugh, "MY2017=2025 GHG Standard for Light Duty Vehicles Mass Reduction," Presentation to the 2013 Great Designs in Steel Conference, Livonia, Michigan, May 2013, available at www.autosteel.org, accessed in August 2013.

4-3. Future Steel Vehicle (FSV) Report Referenced with WorldAutoSteel Council and EDAG Inc., link available at http://www.autosteel.org/, accessed Dec. 2012.

4-4. Singh, H., "Economic Light-Weighting Options for High Volume Production Vehicle Structures for Year 2020," 2012 Great Designs in Steel Conference, Livonia, Michigan, available at www.autosteel.org.

4-5. EDAG Inc., 2012, *Mass Reduction for Light-Duty Vehicles for Model Years 2017–2025*, DOT HS 811 666, U.S. Department of Transportation, National Highway Traffic Safety Administration.

4-6. Smith, T., "Lightweight Steel Wheels Offer Stylish, Low Cost Solutions," presentation to the 2013 Great Designs in Steel Conference, May, 2013, Livonia, Michigan, available at www.autosteel.org, accessed in July 2013.

4-7. Field, F., Kirchain, R., and Roth, R., "Process Cost Modeling: Strategic Engineering and Economic Evaluation of Materials Technologies," *JOM Journal of the Minerals, Metals and Materials Society*, Volume 59, No. 10, 21–32, 2007.

4-8. Shaw, J., "NHTSA's Light Weight Vehicle (LWV) Mult-Material Lightweighting Analysis," 2013 Great Designs in Steel Conference, Livonia, Michigan, available at www.autosteel.org.

Chapter 5
Future Direction of Advanced High-Strength Steels

5.1 Press-Hardened Steel Applications

In this chapter, the future of advanced high-strength steels (AHSS) will be forecasted starting with press-hardened steel (PHS), which is currently the fastest growing area of AHSS, through to second generation AHSS, and then to third generation AHSS. Up until the early 2000s, hot stamping was only used for simple, straight sections such as door beams. However, more recently hot stamping is being used for a large number of complicated body parts where a maximum strength (e.g., 1500 MPa) is desired. In some recently designed vehicles, manufacturers have used as much as 6% PHS on the structural body even though PHS material costs may be as much as 2.4 times higher than other grades of AHSS and may be only about 14% less than aluminum when the extra processing costs of PHS are considered (see estimate in Chapter 4). Besides the higher strength of PHS, the advantages are that very complex parts can be hot stamped because the elongation during stamping is very high and the parts produced can be easily integrated into an existing architecture. A good example of a recently designed vehicle with hot-stamped parts can be seen in which Thomas Grabowski of General Motors Company presented the 2014 Chevrolet Silverado/GMC Sierra Body 1500 Cab Structure Review at the 2013 GDIS [5-1]. Figure 5.1 shows that for the cab-in-white, 6% of the parts are PHS (parts in red). The PHS parts are those that control roof strength and side impact, both of which require very high strength. Also, the B-pillar and the roof rail/A-pillar are typically hard to form. This also favors PHS.

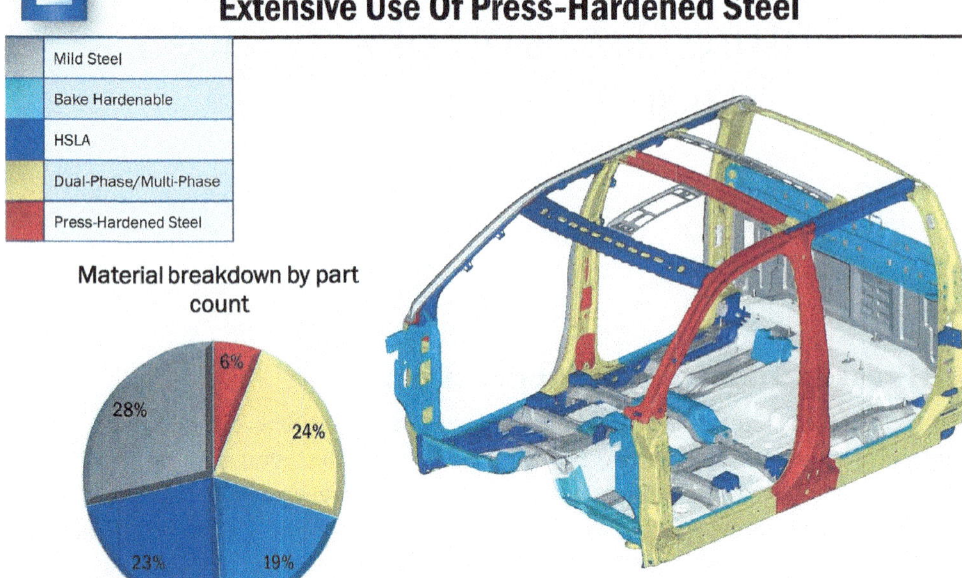

Figure 5.1 PHS parts

Mr. Grabowski also presented the fact that dual cavity tools were used to decrease investment and generate higher volumes since their use on the three cab configurations helped to reduce piece cost (Figure 5.2).

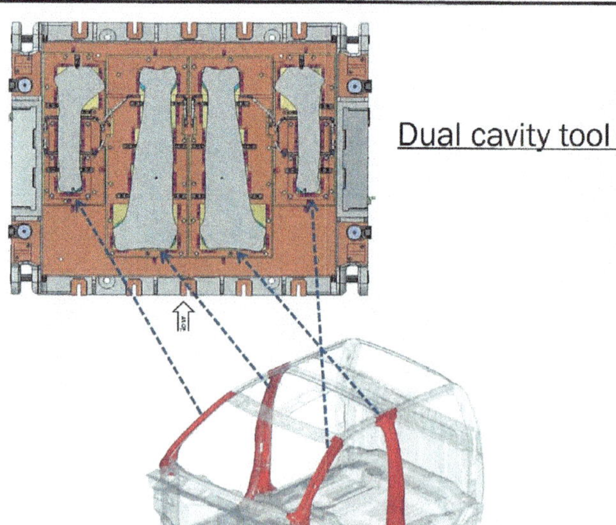

Figure 5.2 Cost reduction

Figure 5.3 shows how the side impact loads are reacted by the PHS members, and Figure 5.4 shows how the roof crush loads are reacted by the PHS parts. In both cases the PHS B-pillar and the roof rail/A-pillar play a big part in reacting these loads.

Reinforcement Safety Cage

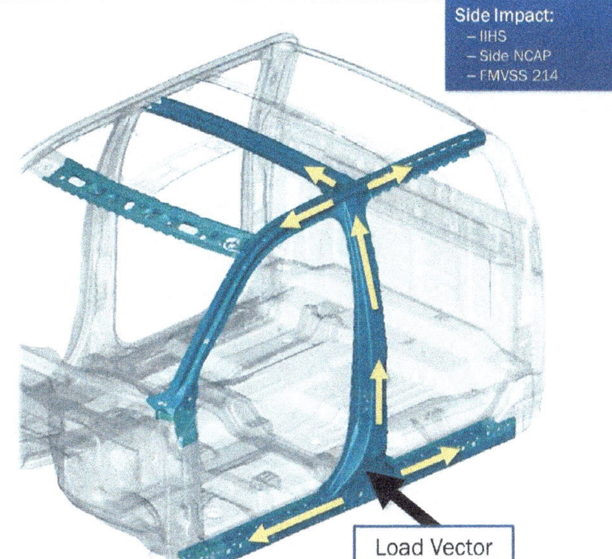

Side impact loads are dissipated through:
- PHS center pillar assembly
- Rocker outer assembly
- PHS roof rail assembly
- #2 PHS roof bow

Figure 5.3 Side impact loads

Reinforcement Safety Cage

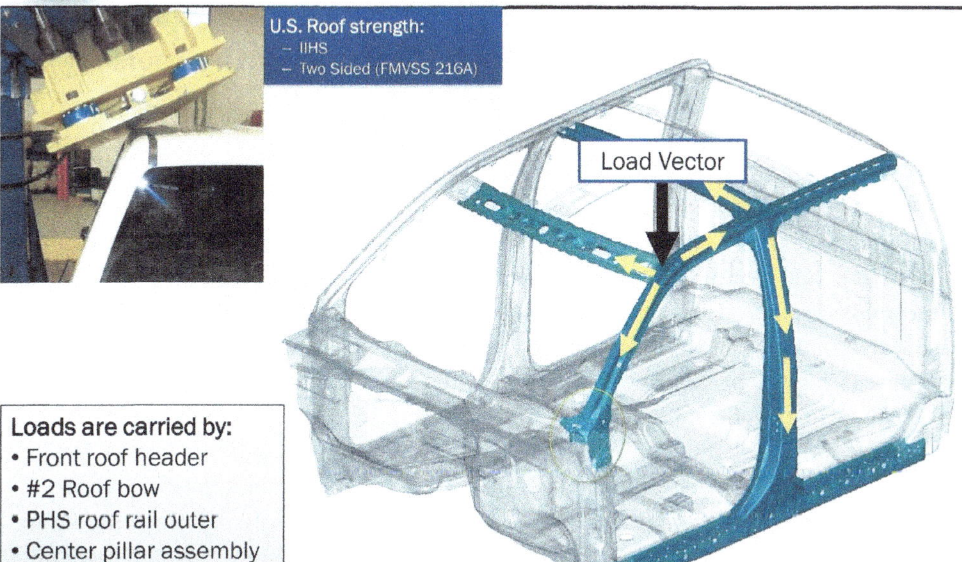

Figure 5.4 Roof crush loads

Figure 5.5 shows how dual-phase steels and PHS were used in conjunction to form the very important B-pillar to rocker joint.

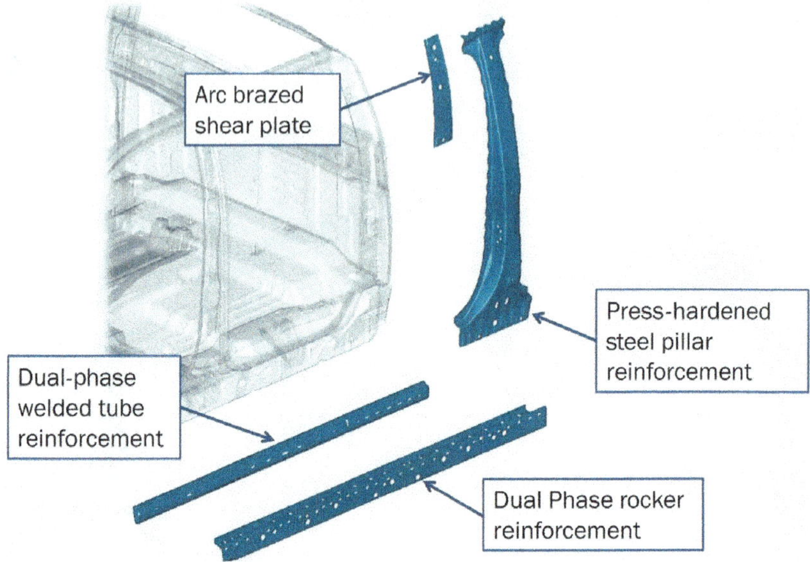

Figure 5.5 B-pillar and rocker components

In summary, the use of PHS stampings on the Silverado enabled a significant improvement in mass efficient safety performance, and the part commonization strategy helped to reduce costs.

5.2 Second Generation Advanced High-Strength Steels

In the next three sections, second and third generation AHSS will be discussed. Figure 5.6 shows the traditional banana chart representing all the common automotive sheet steels. The colored band at the bottom of the graph shows the mild steels, conventional high-strength steels (CHSS), and the first generation AHSS. Hot-stamped boron (HSB) is often considered to be a first generation AHSS. At the top of the chart, balloons are shown for austenitic stainless steel and twinning-induced-plasticity (TWIP) steels. These steels are considered to be second generation AHSS. Lightweight induced plasticity (L-IP) is also considered to be a second generation AHSS but is not commercially available. The second generation AHSS provide very high elongation at ultra-high strength when compared with first generation AHSS. Given that TWIP and austenitic stainless are quite expensive and production issues have stalled implementation of TWIP, these steels have not made significant inroads into the automotive market. Austenitic stainless steels have penetrated the appliance industry by providing premium exteriors and high formability for the surfaces and handles of those products. However, this is at a cost essentially equivalent to aluminum. This explains why, like aluminum, austenitic stainless also has not made major inroads into the automotive market. Exhaust systems use stainless for corrosion purposes but use a less formable grade than austenitic stainless steels and are therefore not suited for most structural automotive components. Many steel experts believe that the extreme elongation provided by the second generation AHSSs is not required from a formability perspective. They (and I) believe that if someone could develop a steel that had the same strength and half the cost of second generation AHSS and with an elongation between first generation and second generation AHSS, that this combination of attributes would be all that is required to minimize weight.

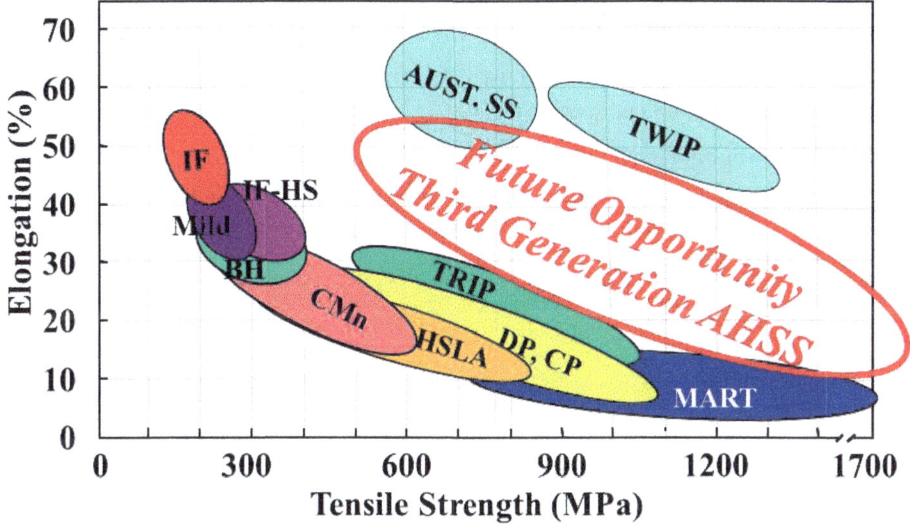

Figure 5.6 Steel banana chart

Different strategies of producing third generation AHSS will be discussed in the next section.

5.3 Third Generation Advanced High-Strength Steels

There are a number of strategies for making third generation AHSS (Figure 5.7). These include the following:

1. Enhanced dual-phase steel
2. Modified transformation-induced plasticity (TRIP)
3. Ultrafine bainite
4. Quenching and partitioning (Q&P)
5. Lower manganese TRIP/TWIP
6. Higher manganese TRIP
7. TRIP-aided bainite-ferrite (TBF)

8. Carbide-free bainite (CFB)
9. NanoSteel AHSS

None of these steels are commercially available as yet but are being extensively pursued by a number of organizations. A discussion of some of these strategies for producing third generation AHSSs is in Matlock and Spear [5-2]. This book will not discuss the strategies for making third generation AHSS but will focus on properties and applications of the NanoSteel steels. The following are reasons for discussing NanoSteel's third generation AHSS versus the other strategies:

1. NanoSteel is already into production trials for its steels.
2. There has been considerable research (which is in the public domain) as to the weight benefits of using NanoSteel's products.
3. In the author's opinion, one or more of the third generation AHSSs will be available in the future, and the mechanical properties may be close to the NanoSteel products. Therefore, NanoSteel products are probably a good proxy for demonstrating the future of AHSS.

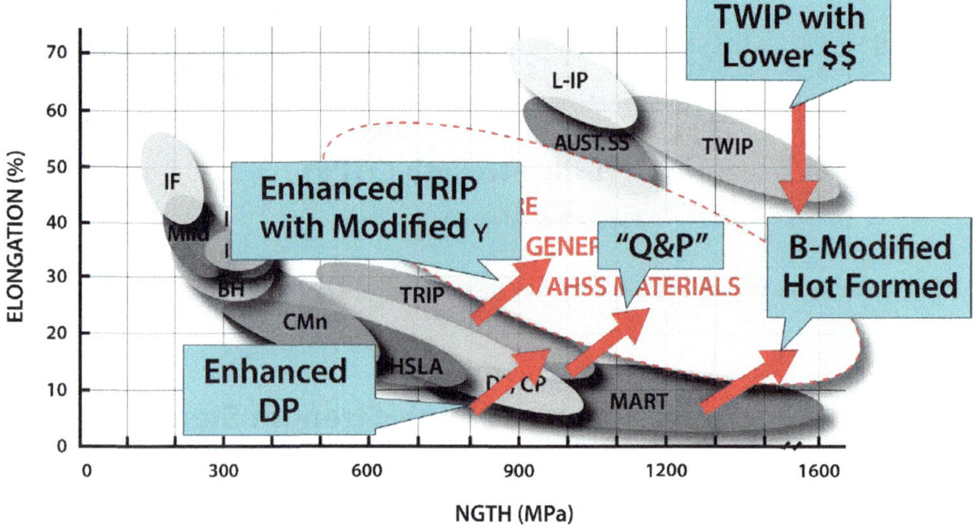

Figure 5.7 Third generation AHSS development

Therefore, Dr. Daniel Branagan's presentation from the 2013 Great Designs in Steel Conference (GDIS) will be reviewed [5-3]. In his presentation, Dr. Branagan discussed in detail the third generation AHSS that his company (the NanoSteel Company, Inc.) is developing and the key differentiators of the company's approach.

The NanoSteel Company was founded in 2002 (Figure 5.8), but their discovery of nano-structured steels goes back to 1996. Early attempts at producing nano-structured steels focused on metallic coatings and thin foil technology. More recently, NanoSteel has been developing automotive sheet metal and started industrial production trials in 2012.

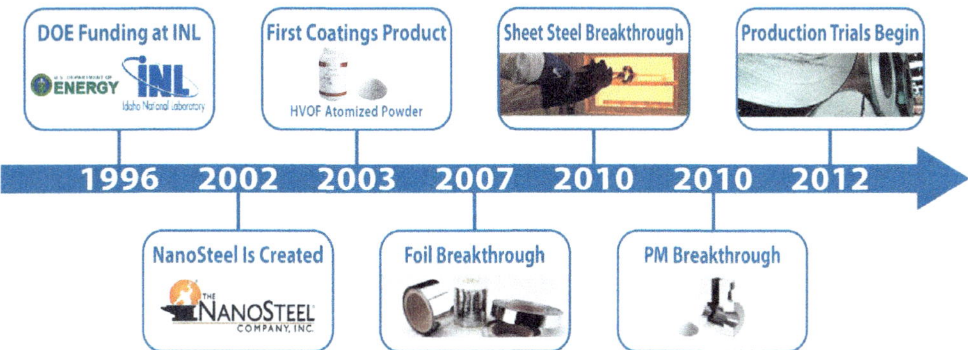

Figure 5.8 History of NanoSteel Company

NanoSteel's approach to developing third generation AHSS resulted from a different development strategy than the ones referenced above (Figure 5.9) with NanoSteel graphed as elongation versus tensile strength (UTS). The key differentiators of NanoSteel's approach are chemistry, structural formation, and tensile behavior. Regarding chemistry, NanoSteel uses greater percentages of P-group elements and in different ratios than have previously been utilized (Figure 5.10). This strategy accounts for the dramatic increases in strength and was derived from NanoSteel's experience with metallic coating and foil technologies.

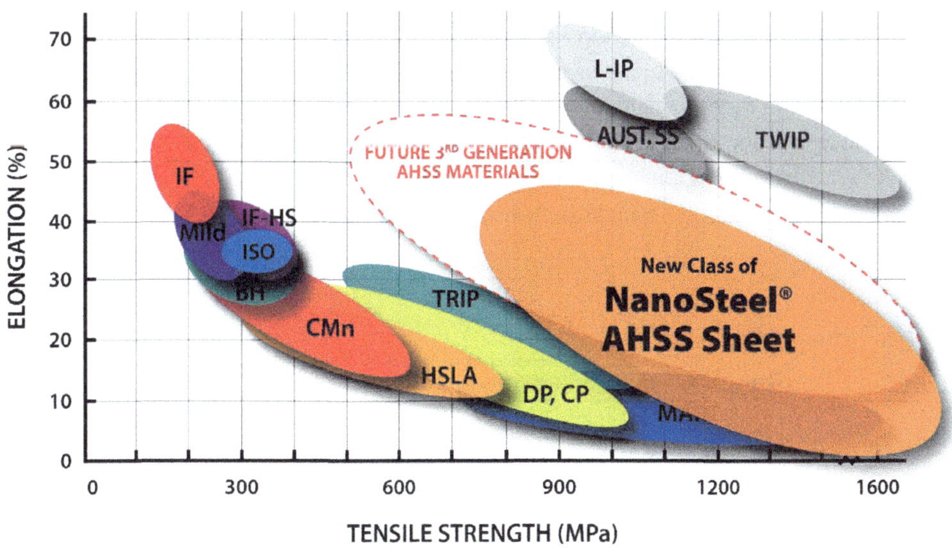

Figure 5.9 Third generation AHSS: NanoSteel (NS)

Periodic Table of Elements

P-Group Elements

Iron

hydrogen 1 H 1.0079																		helium 2 He 4.0026
lithium 3 Li 6.941	beryllium 4 Be 9.0122												boron 5 B 10.811	carbon 6 C 12.011	nitrogen 7 N 14.007	oxygen 8 O 15.999	fluorine 9 F 18.998	neon 10 Ne 20.180
sodium 11 Na 22.990	magnesium 12 Mg 24.305												aluminium 13 Al 26.982	silicon 14 Si 28.086	phosphorus 15 P 30.974	sulfur 16 S 32.065	chlorine 17 Cl 35.453	argon 18 Ar 39.948
potassium 19 K 39.098	calcium 20 Ca 40.078		scandium 21 Sc 44.956	titanium 22 Ti 47.867	vanadium 23 V 50.942	chromium 24 Cr 51.996	manganese 25 Mn 54.938	iron 26 Fe 55.845	cobalt 27 Co 58.933	nickel 28 Ni 58.693	copper 29 Cu 63.546	zinc 30 Zn 65.39	gallium 31 Ga 69.723	germanium 32 Ge 72.61	arsenic 33 As 74.922	selenium 34 Se 78.96	bromine 35 Br 79.904	krypton 36 Kr 83.80
rubidium 37 Rb 85.468	strontium 38 Sr 87.62		yttrium 39 Y 88.906	zirconium 40 Zr 91.224	niobium 41 Nb 92.906	molybdenum 42 Mo 95.94	technetium 43 Tc [98]	ruthenium 44 Ru 101.07	rhodium 45 Rh 102.91	palladium 46 Pd 106.42	silver 47 Ag 107.87	cadmium 48 Cd 112.41	indium 49 In 114.82	tin 50 Sn 118.71	antimony 51 Sb 121.76	tellurium 52 Te 127.60	iodine 53 I 126.90	xenon 54 Xe 131.29
caesium 55 Cs 132.91	barium 56 Ba 137.33	57-70 *	lutetium 71 Lu 174.97	hafnium 72 Hf 178.49	tantalum 73 Ta 180.95	tungsten 74 W 183.84	rhenium 75 Re 186.21	osmium 76 Os 190.23	iridium 77 Ir 192.22	platinum 78 Pt 195.08	gold 79 Au 196.97	mercury 80 Hg 200.59	thallium 81 Tl 204.38	lead 82 Pb 207.2	bismuth 83 Bi 208.98	polonium 84 Po [209]	astatine 85 At [210]	radon 86 Rn [222]
francium 87 Fr [223]	radium 88 Ra [226]	89-102 * *	lawrencium 103 Lr [262]	rutherfordium 104 Rf [261]	dubnium 105 Db [262]	seaborgium 106 Sg [266]	bohrium 107 Bh [264]	hassium 108 Hs [269]	meitnerium 109 Mt [268]	ununnilium 110 Uun [271]	unununium 111 Uuu [272]	ununbium 112 Uub [277]		ununquadium 114 Uuq [289]				

Series														
Lanthanide Series	lanthanum 57 La 138.91	cerium 58 Ce 140.12	praseodymium 59 Pr 140.91	neodymium 60 Nd 144.24	promethium 61 Pm [145]	samarium 62 Sm 150.36	europium 63 Eu 151.96	gadolinium 64 Gd 157.25	terbium 65 Tb 158.93	dysprosium 66 Dy 162.50	holmium 67 Ho 164.93	erbium 68 Er 167.26	thulium 69 Tm 168.93	ytterbium 70 Yb 173.04
Actinide Series	actinium 89 Ac [227]	thorium 90 Th 232.04	protactinium 91 Pa 231.04	uranium 92 U 238.03	neptunium 93 Np [237]	plutonium 94 Pu [244]	americium 95 Am [243]	curium 96 Cm [247]	berkelium 97 Bk [247]	californium 98 Cf [251]	einsteinium 99 Es [252]	fermium 100 Fm [257]	mendelevium 101 Md [258]	nobelium 102 No [259]

Figure 5.10 P-Group elements in NanoSteel products

To develop automotive products greater than .5 mm, NanoSteel did not need the extreme strength of the foils and metallic coatings products, but much more ductility. Figure 5.11 shows a comparison of the automotive target elongation and strength versus the thinner products. To achieve the required increase in ductility and the formation of specific characteristic nanoscaled structures, NanoSteel is using a lower atomic percentage of P-group elements than the previous generations of metallic coating and foil technology (Figure 5.12), but still much higher than for first generation AHSS (Figure 5.13). Figure 5.14 shows the atomic percentage of P-group elements for the NanoSteel third generation steels compared with the first generation AHSS and the earlier NanoSteel products.

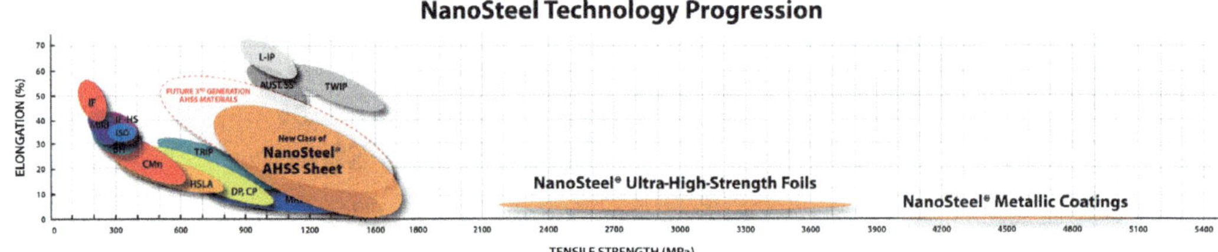

Figure 5.11 NanoSteel technology progression

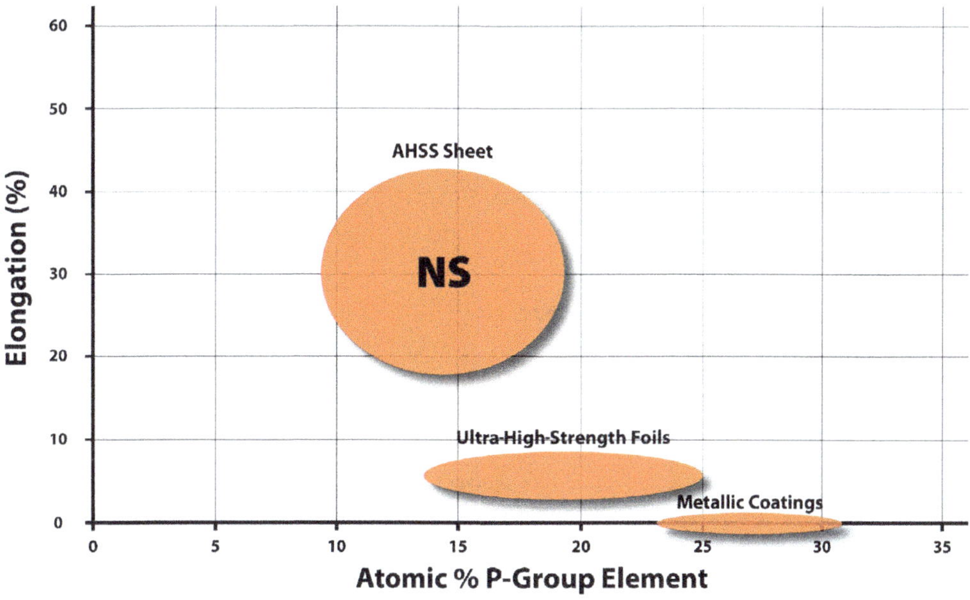

Figure 5.12 NanoSteel technology; % P-group element to elongation

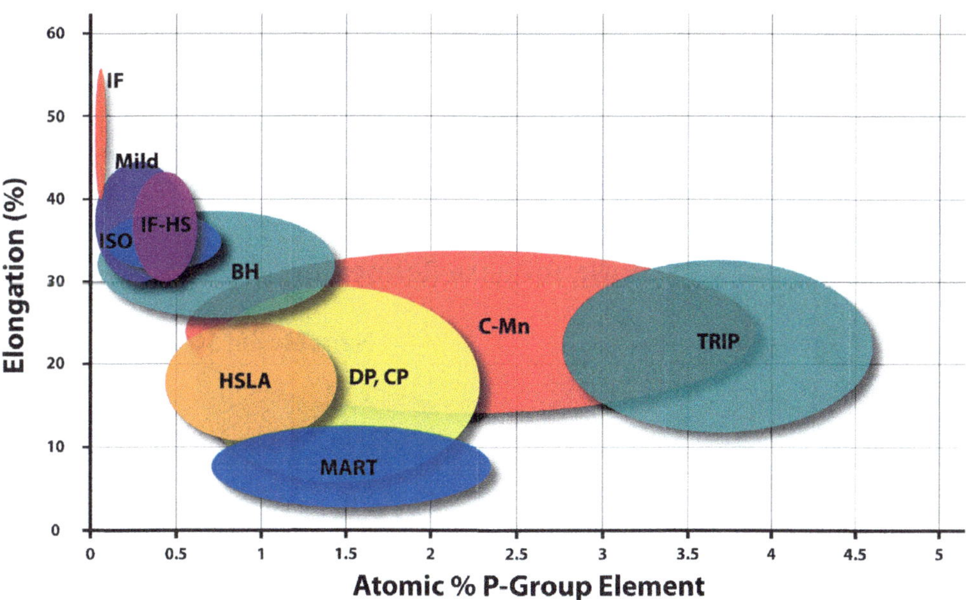

Figure 5.13 First generation AHSS; % P-group element to elongation

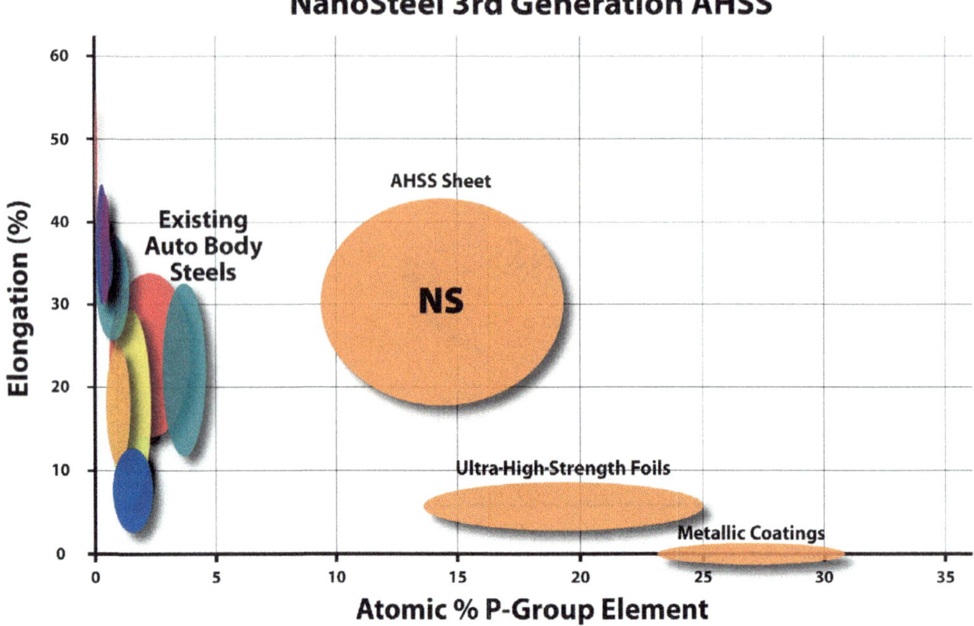

Figure 5.14 NanoSteel (NS) versus first generation AHSS; % P-group element to elongation

NanoSteel's third generation AHSS is also distinct in its structural formation and the process by which it is accomplished. The reaction of NanoSteel's materials to heat treatment gives rise to a refinement in grain size, reducing the size of both the matrix and pinning phases, which contributes to the strength and ductility of the steel (Figure 5.15). There is no quenching step required in the heat treatment.

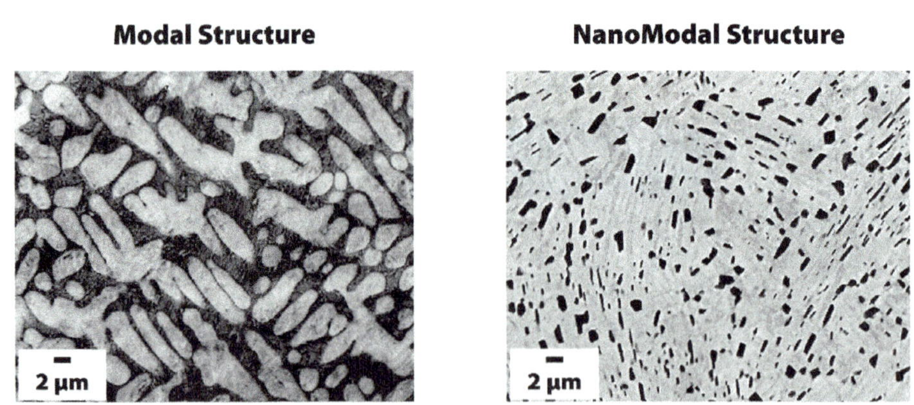

Figure 5.15 Static NanoPhase refinement

Early development of the company's automotive sheet steels showed that NanoSteel's AHSS could be made in three different classes, which covered the full range of the third generation AHSS target (Figure 5.16). In Figure 5.17 the laboratory samples of the three different grades of NanoSteel AHSS are compared with several first generation AHSS using engineering stress-strain characteristics. Production trials of NanoSteel's automotive AHSS are under way at the time of this writing.

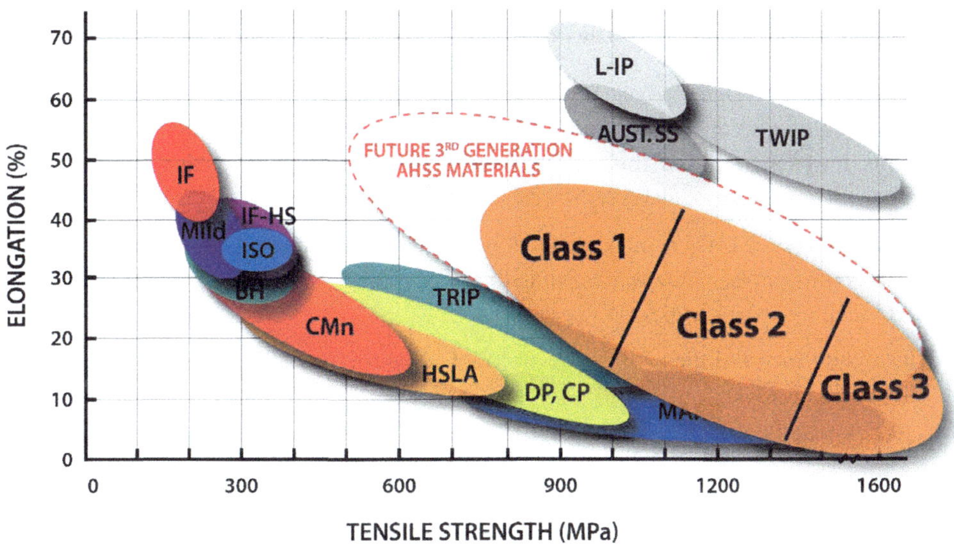

Figure 5.16 Tensile behavior: NanoSteel

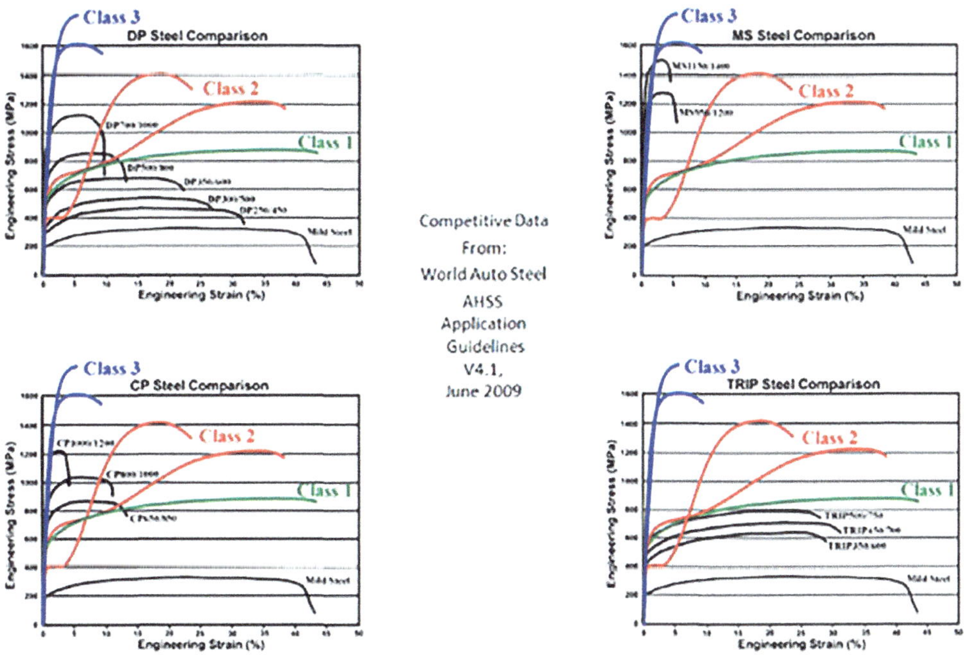

Figure 5.17 Tensile behavior: comparison with existing AHSS

5.4 Ultimate Weight Benefits of Advanced High-Strength Steels

After EDAG/ National Highway & Traffic Administration Safety (NHTSA) completed the 2011 Honda Accord study (which was covered in Chapter 4) NanoSteel contracted with Harry Singh at EDAG Engineering to extend the study using NanoSteel materials (Figure 5.18). The complete report of this study can be found on NanoSteel's website, www.nanosteelco.com/EDAG. In this section, the salient points of this study will be covered with the permission of NanoSteel. The primary assumptions of the study were as follows:

1. Though the lightweight vehicle (LWV) study was a full vehicle study, the NanoSteel study would only consider the body-in-white without the closures, glass, and bumper beams. It was thought that this system of the vehicle provided the greatest opportunity for conversion to NanoSteel AHSS.
2. The version of the LWV using NanoSteel AHSS needed to have the same vehicle attributes (e.g., noise, vibration, and harshness [NVH] and safety performance) as the 2011 Honda Accord and the LWV.
3. The architecture and the joining (e.g., welding and structural adhesives) was identical to the LWV.
4. All but about 3% of the steels in the LWV were replaced with NanoSteels N1, N2, and N3, with progressively higher tensile strengths. The 3% of the body structure not replaced with NanoSteel's AHSS was in small brackets. A table showing the replacement strategy is shown in Figure 5.19. The groupings were derived from the elongations of N1, N2, and N3 compared with the steels being replaced in order to both have the ability to form the parts as previously designed for the LWV and to ensure that the vehicle would have similar crash modes. All of the ultra-high-strength steels from the LWV were replaced with N3, the other AHSS grades were replaced with N2, and the mild and CHSS grades were replaced with N1. Also DP 300/500, which is usually considered to be an AHSS, was replaced with N1. A visual representation of the replacement strategy is shown in Figure 5.20.
5. Formability of the geometries used in the LWV design was validated during the NHTSA study with one-step forming simulations utilizing the original AHSS grades. Formability analysis was not part of the NanoSteel study, but the elongations of N1, N2, and N3 probably justified the assumption that they would be formable for the panels for which they were used. However, N1, N2, and N3 may lead to more springback than for the steels being replaced. Full formability assessment of NanoSteel's AHSS will take place with the release of validation materials to the manufacturers.
6. Costs were not considered as part of this study. For instance, the mild steels in the LWV were replaced with N1. Although this was acceptable from a formability perspective, this substitution may not have been appropriate if a full cost optimization process had been followed. Also, the manufacture of third generation AHSS may present surface quality problems for Class A exposed surfaces.

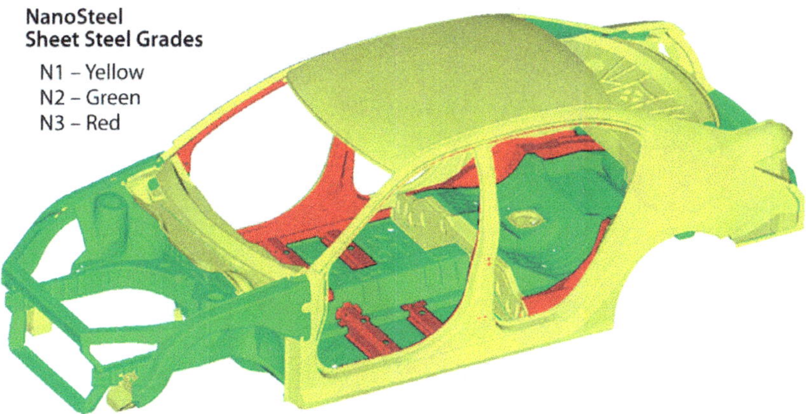

Figure 5.18 NanoSteel intensive body-in-white

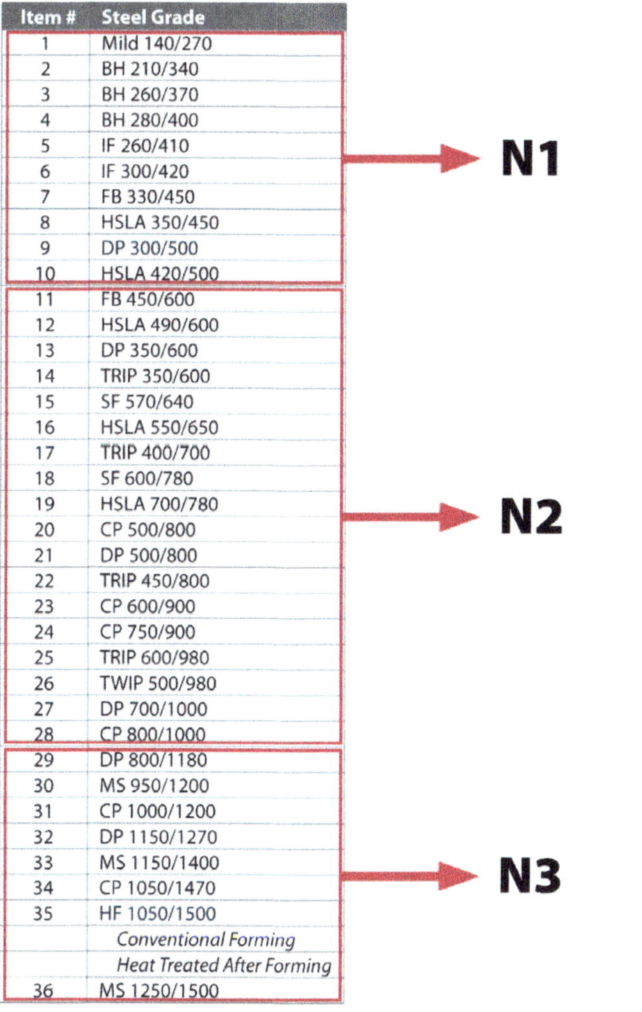

Item #	Steel Grade	
1	Mild 140/270	N1
2	BH 210/340	N1
3	BH 260/370	N1
4	BH 280/400	N1
5	IF 260/410	N1
6	IF 300/420	N1
7	FB 330/450	N1
8	HSLA 350/450	N1
9	DP 300/500	N1
10	HSLA 420/500	N1
11	FB 450/600	N2
12	HSLA 490/600	N2
13	DP 350/600	N2
14	TRIP 350/600	N2
15	SF 570/640	N2
16	HSLA 550/650	N2
17	TRIP 400/700	N2
18	SF 600/780	N2
19	HSLA 700/780	N2
20	CP 500/800	N2
21	DP 500/800	N2
22	TRIP 450/800	N2
23	CP 600/900	N2
24	CP 750/900	N2
25	TRIP 600/980	N2
26	TWIP 500/980	N2
27	DP 700/1000	N2
28	CP 800/1000	N2
29	DP 800/1180	N3
30	MS 950/1200	N3
31	CP 1000/1200	N3
32	DP 1150/1270	N3
33	MS 1150/1400	N3
34	CP 1050/1470	N3
35	HF 1050/1500	N3
	Conventional Forming	N3
	Heat Treated After Forming	N3
36	MS 1250/1500	N3

Figure 5.19 Material selection for NanoSteel body-in-white

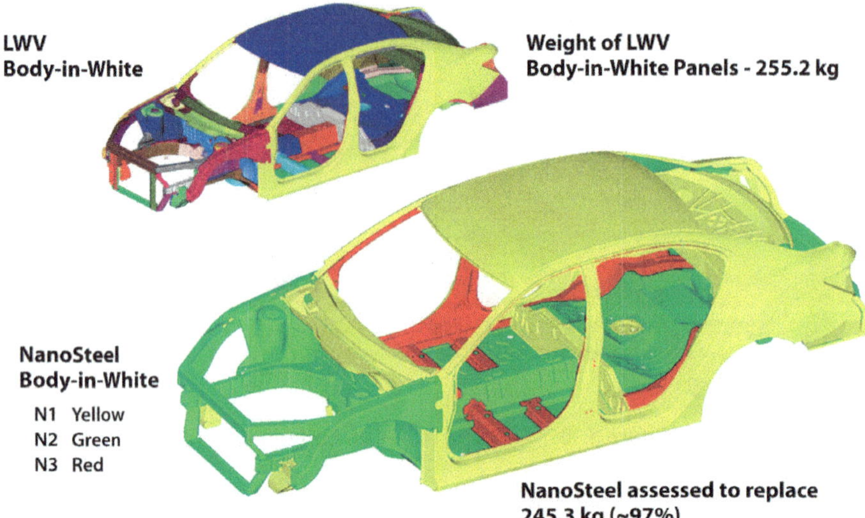

Figure 5.20 Visualization of NanoSteel material selection for body-in-white

A flowchart of the process, which EDAG used to optimize the body-in-white, is shown in Figure 5.21. The process consisted of the following major steps:

1. NanoSteel supplied EDAG with the properties of N1, N2, and N3, including stress/strain curves and strain rate sensitivity.
2. NanoSteel materials were then chosen to replace the LWV panels as discussed previously.
3. Gauges were selected for the different panels using the thicknesses of the LWV panels. Typically, the starting thicknesses for the NanoSteel AHSS were about 10% less than the LWV steels, but not in all cases. Judgment was used when it was obvious that 10% less would create a problem (e.g., when the panels were already very thin). Starting with a low minimum thickness of 0.5 mm, the optimization in most cases pushed the minimum thickness higher.
4. Next, the LWV finite-element NASTRAN model was adjusted to account for the new gauges and run for an initial set of NVH properties. Bending and torsion static displacements and various critical normal modes were calculated using the NASTRAN model.
5. The GENESIS optimizer was used to optimize the thicknesses to get as close as possible to the 2011 Honda Accord target test values for the stiffnesses and normal modes, acknowledging that these values would be further enhanced through changes made to meet the crash standards.
6. The material and thickness values and the strain rate sensitivity data were then put into the LS-Dyna model for the LWV to assess crash worthiness. Using the LS-Dyna crash model and an optimization tool called HEEDS (Red Cedar Technologies, Inc.), the most critical load cases were used to optimize the thicknesses:
   a. Frontal impact with flat rigid wall barrier (US NCAP).

b. IIHS frontal impact with offset deformable barrier.

c. FMVSS side pole impact.

d. IIHS roof crush.

e. Two other crash load cases were run after the optimization to confirm acceptable performance: The IIHS side impact with a moveable deformable barrier and the FMVSS No. 301 rear impact test for fuel tank integrity.

7. The final thickness parameters for the NanoSteel body were compared with the Honda test results for all the critical crash and NVH targets. If acceptable, the weight of the NanoSteel body-in-white was computed and compared with the Honda weight and the LWV weight.

The results of the initial NanoSteel linear run along with the results after GENESIS optimization are shown in Figures 5.22 and 5.23. In general, the linear values were much too low in the original run but came into an acceptable range after optimization. Though the torsional stiffness was a bit low, it was judged to be close enough to the Honda Accord target to proceed to the next step. Typically, after crash optimization the linear results go up.

Vehicle	Torsion Stiffness (kN-m/deg)	Bending Stiffness (N/mm)
2011 Honda Accord Test (Target)	12.33	8,690
LWV	14.40	11,760
NanoSteel Initial Run	10.73	8,366
NanoSteel Genesis Optimization	10.49	8,510

Figure 5.22 GENESIS optimization NanoSteel body-in-white bending and torsional stiffness

Vehicle	Front-End Lateral Mode (Hz)	Second Order Bending Mode (Hz)	First Order Bending Mode (Hz)	Torsion Mode (Hz)
2011 Honda Accord Test Frequency (Target)	35.1	39.3	44.2	50.1
LWV Modeled Frequency	40.5	40.8	46.3	48.7
NanoSteel Initial Run Modeled Frequency	34.2	36.6	40.8	45.5
NanoSteel Genesis Optimized Modeled Frequency	35.7	39.0	43.5	48.5

Figure 5.23 GENESIS optimization NanoSteel body-in-white normal modes

For crash analysis and optimization, NanoSteel supplied the nonlinear material properties for N1, N2, and N3. The table in Figure 5.24 shows the static properties, and the curves in Figures 5.25, 5.26, and 5.27 show the true stress-strain data along with the strain rate sensitivity data.

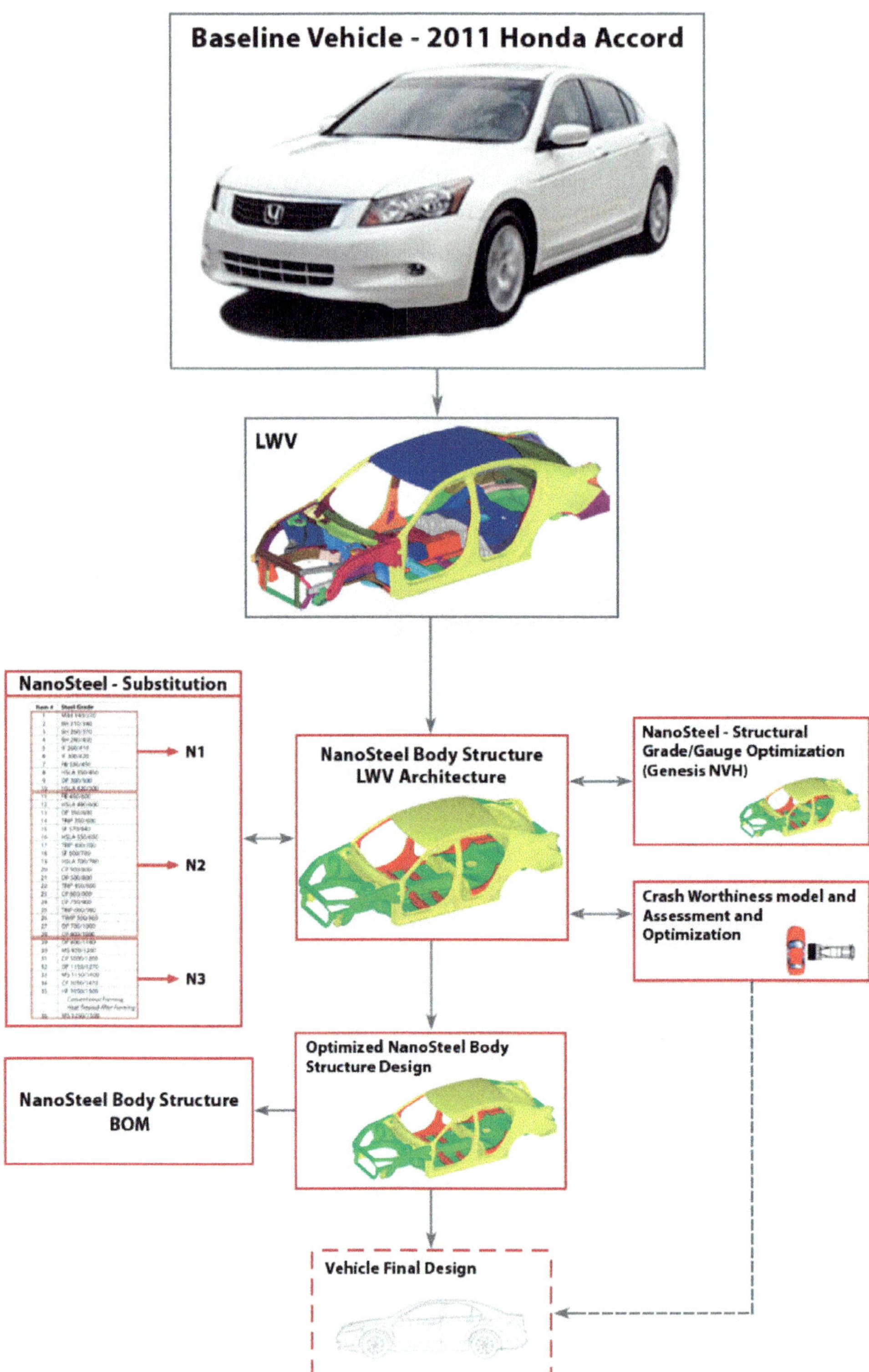

Figure 5.21 NanoSteel body-in-white flowchart

NanoSteel Grade	Yield (MPa)	Ultimate Tensile (MPa)	Elongation (%)
N1	450	950	35%
N2	450	1200	20%
N3	1000	1600	12%

Figure 5.24 NanoSteel static material properties

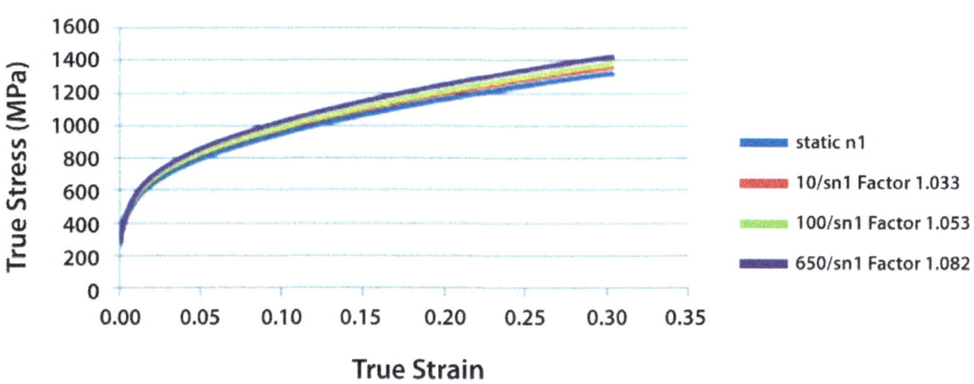

Figure 5.25 NanoSteel N1—True stress/strain curves

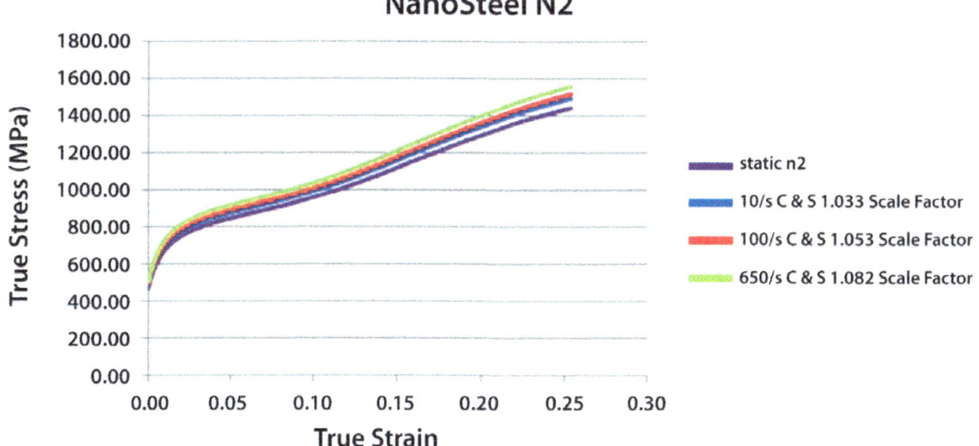

Figure 5.26 NanoSteel N2—True stress/strain curves

Stress vs. Strain Cowper & Symonds, Test Factors

NanoSteel N3

True Stress (MPa)

2000
1500
1000
500
0

0.00 0.02 0.04 0.06 0.08 0.10 0.12

True Strain

static n3
10/sn3 Factor 1.033
100/sn3 Factor 1.053
650/sn3 Factor 1.082

Figure 5.27 NanoSteel N3—True stress/strain curves

The results of the final crash optimization, which resulted in the final thickness values, will now be reviewed. Figure 5.28 shows a comparison of the 2011 Honda Accord, the LWV, and the optimized NanoSteel vehicle for the full frontal NCAP Crash. The Honda Accord, the LWV, and the NanoSteel vehicle all had approximately the same crash pulse, which would indicate that the driver and passenger would have essentially the same head index criteria (HIC) numbers. Similarly, for the 40% offset barrier test, all three vehicles had the same deceleration pulse (Figure 5.29). However, in this case, Honda Crosstour data were used because the Accord data were unavailable and the Crosstour platform very closely resembles that of the Accord. Figure 5.30 shows that the occupant compartment intrusion ratings of all three vehicles were in the good range for the 40% offset test.

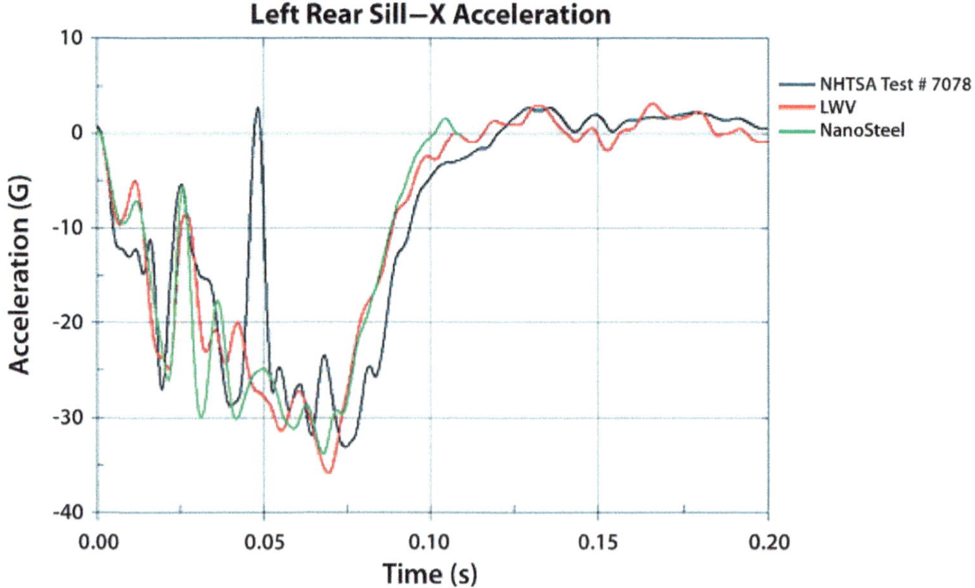

Figure 5.28 Frontal NCAP crash pulse—2011 Honda Accord, LWV, and NanoSteel vehicles

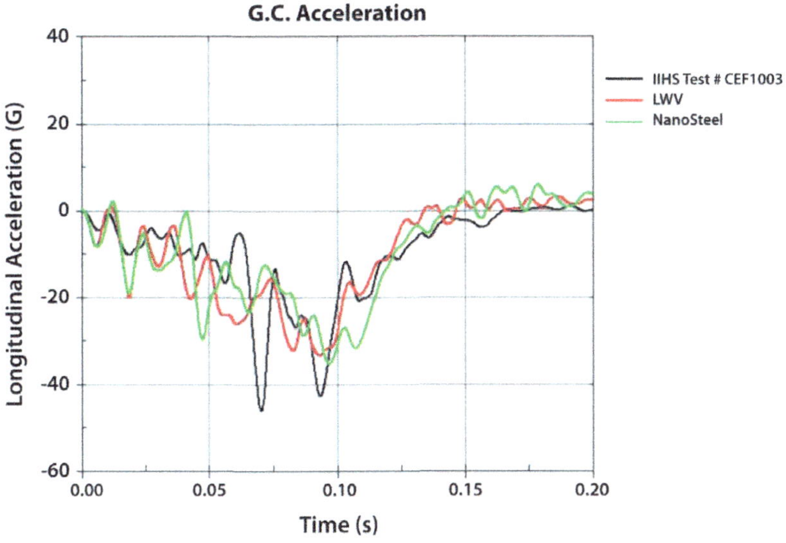

Figure 5.29 IIHS frontal crash pulse for the 2011 Honda Crosstour, LWV, and NanoSteel vehicles

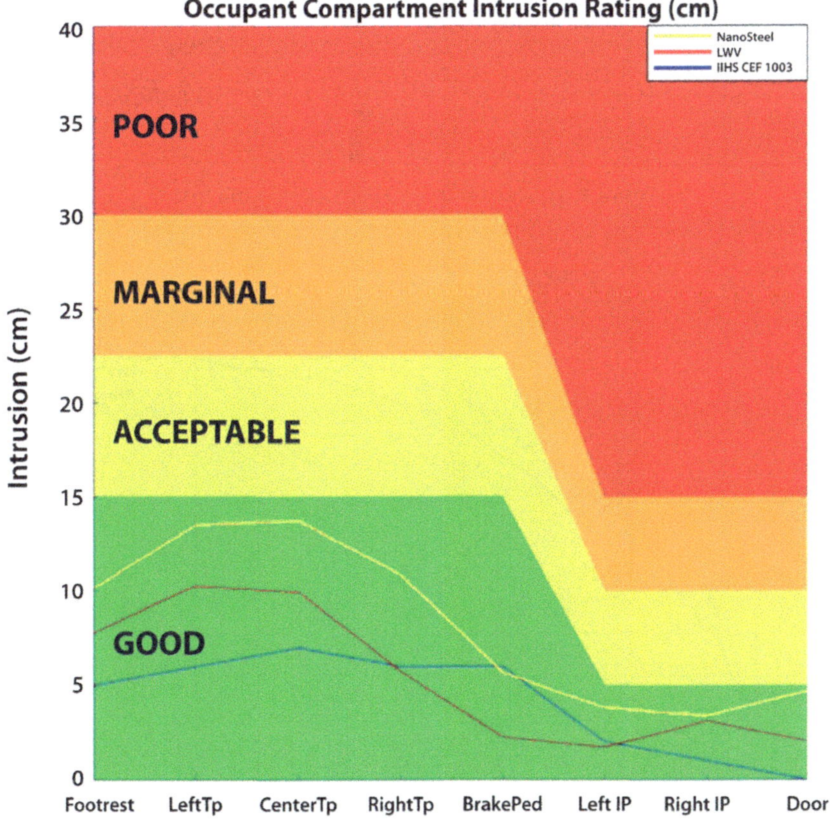

Figure 5.30 40% offset intrusions for the 2011 Honda Crosstour, LWV, and NanoSteel vehicles

Another crash test for which the thicknesses were optimized was the side pole test. The configuration for this test is shown in Figure 5.31. The mid B-pillar velocities for the three vehicles are shown in Figure 5.32, and the exterior crush at the H-point level is shown in Figure 5.33 using the final thickness values for the NanoSteel vehicle. The results in Figures 5.32 and 5.33 show all three vehicles to be about the same.

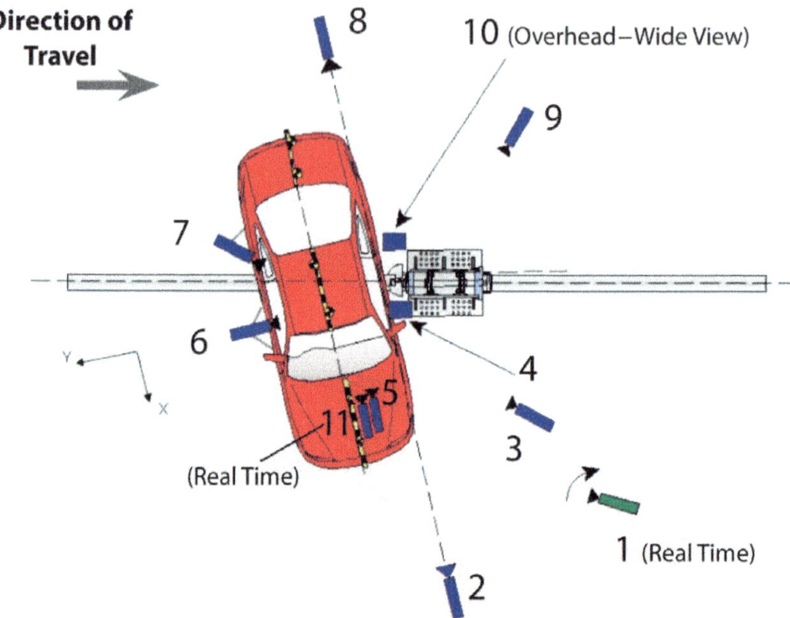

Figure 5.31 Test setup for FMVSS Side pole test

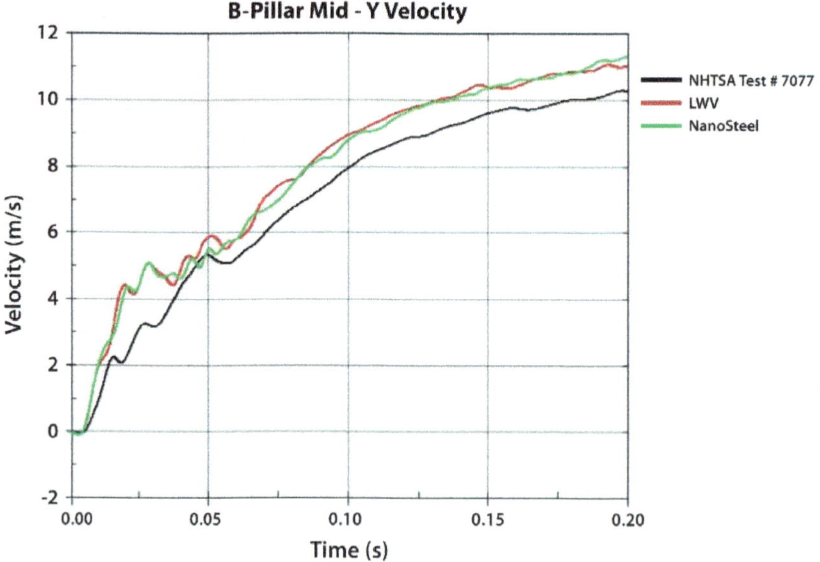

Figure 5.32 Lateral velocity versus time for the mid-B-pillar

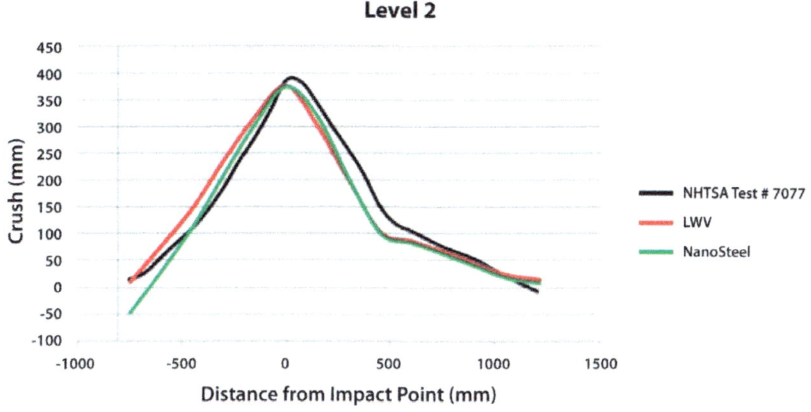

Figure 5.33 Exterior crush at the H-point level for the side pole test

The configuration for the IIHS deformable barrier test is shown in Figure 5.34. The velocity results are shown in Figure 5.35, and the deformation results are shown in Figure 5.36. Again the results are about the same for all three vehicles.

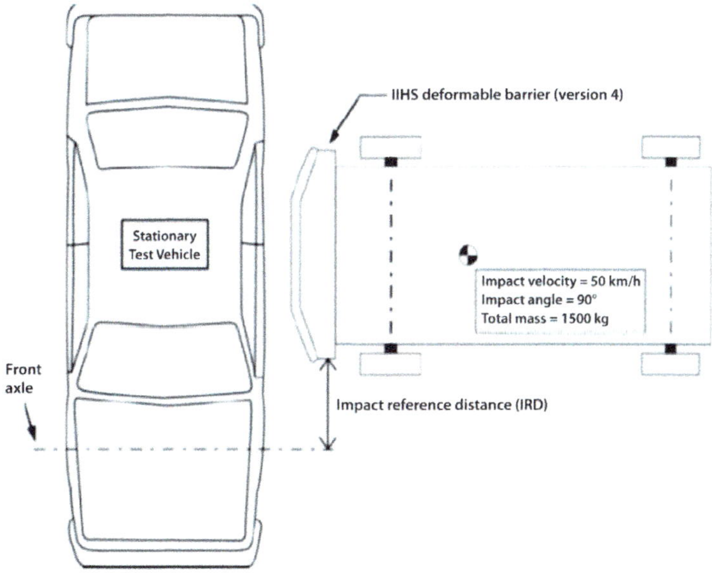

Figure 5.34 IIHS moving deformable barrier test

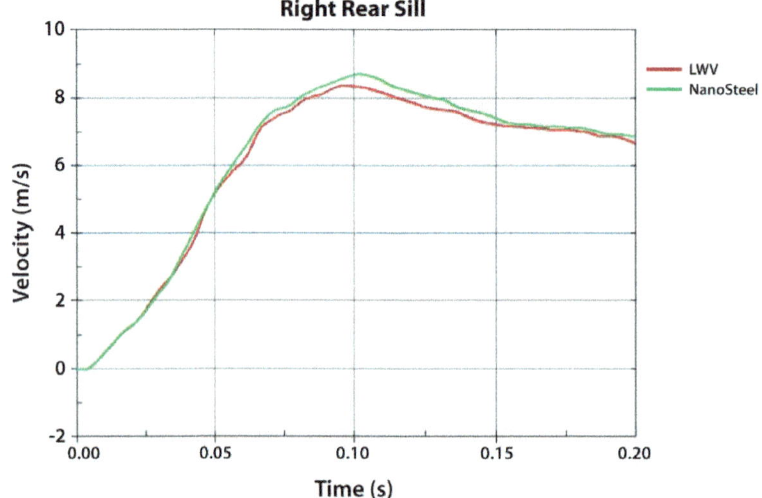

Figure 5.35 Right rear sill velocity for the IIHS deformable barrier test

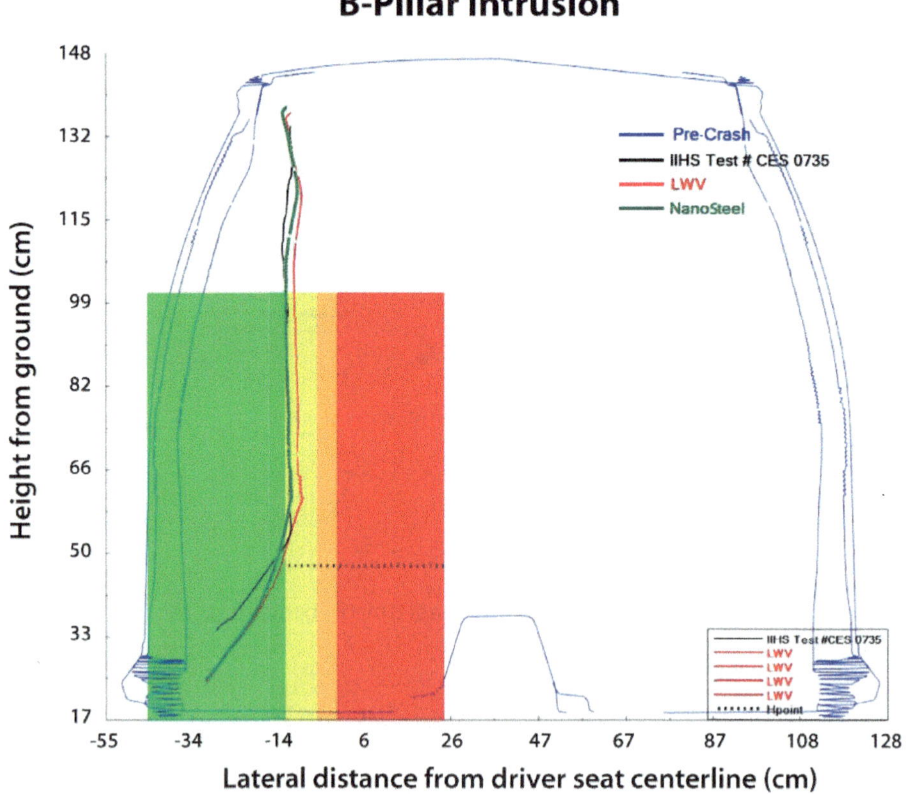

Figure 5.36 Intrusion for the IIHS deformable barrier test

The configuration for the IIHS roof crush test is shown in Figure 5.37, and the plate intrusion results are shown in Figure 5.38. In this test, the NanoSteel vehicle outperformed the LWV design and Honda Accord.

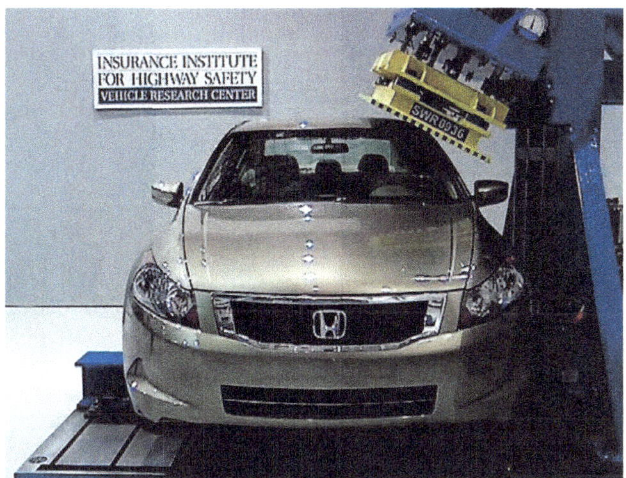

Figure 5.37 Test set-up for the IIHS roof crush test

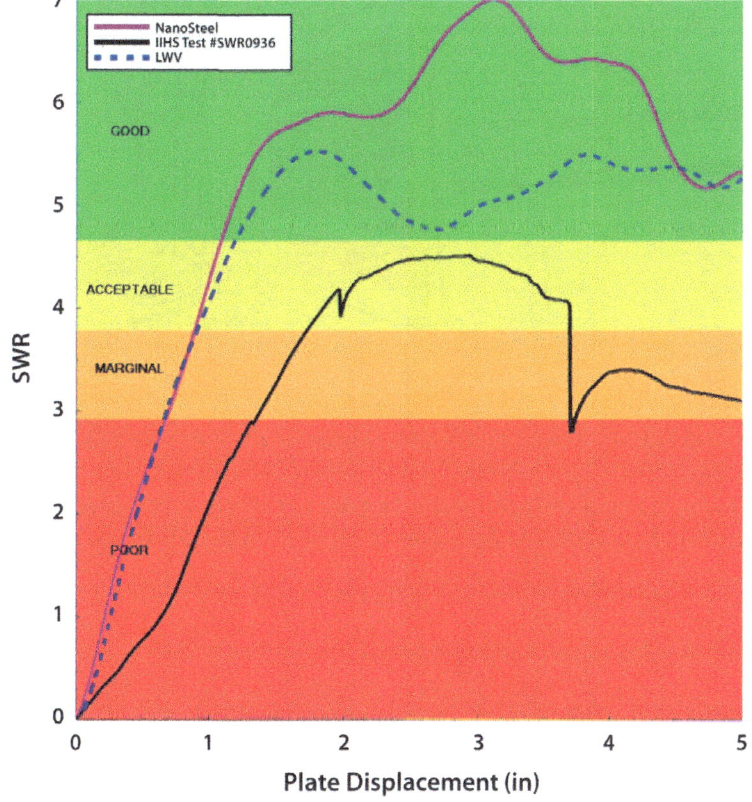

Figure 5.38 SWR-Platen force divided by curb weight for IIHS roof crush

The configuration for FMVSS No. 301, which is a test for gas tank fuel leakage, is shown in Figure 5.39, and a view of the NanoSteel crash model after the test is shown in Figure 5.40. The picture in Figure 5.40 shows no deformation of the fuel tank, demonstrating no leakage would occur.

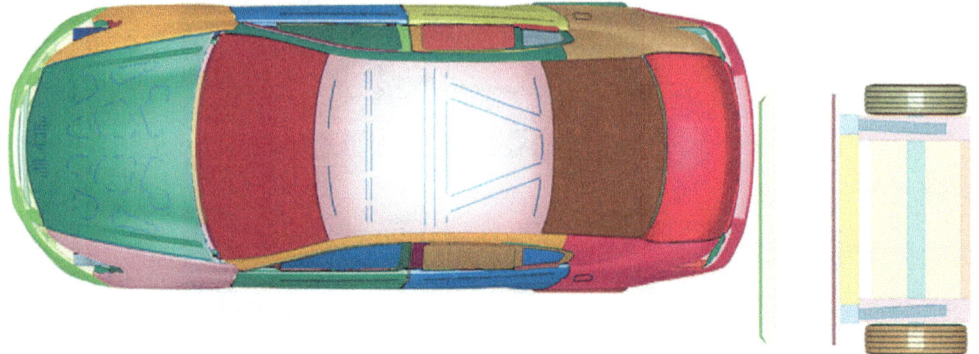

Figure 5.39 Test setup for FMVSS No. 301 test

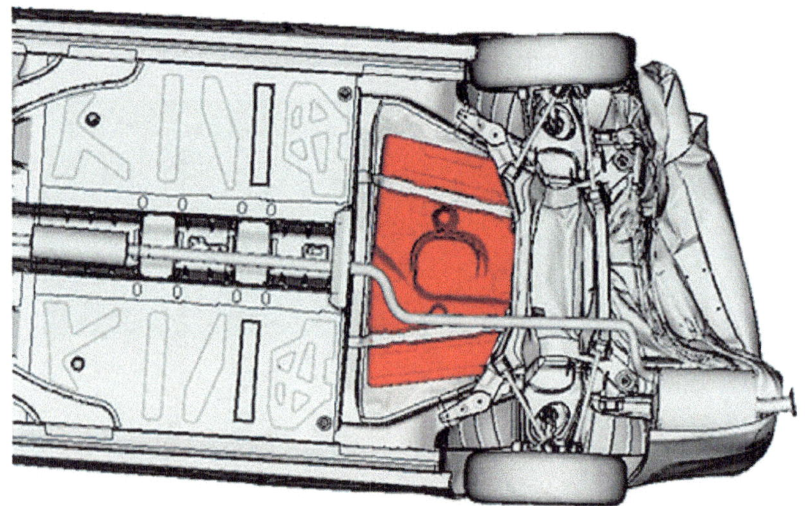

Figure 5.40 Bottom view of the NanoSteel body-in-white after FMVSS No. 301 test

A summary of the crash results is shown in the table in Figure 5.41. Overall, the NanoSteel vehicle performed at an equivalent level to the Honda Accord and LWV design.

Structural Response of the NanoSteel Body Structure		
Test	**Dynamic**	**Static**
NCAP Frontal	Acceleration magnitude and the pulse time width are similar to the baseline Honda Accord. For airbag deployment, the average acceleration during 0.005 to 0.015 seconds is 8.5 G's	Comparable to baseline Honda Accord and LWV
IIHS 40% Offset Frontal	Acceleration about the same magnitude as the Honda Crosstour, and the pulse time width is about the same as the pulse width of the Honda Crosstour	Comparable to the Honda Crosstour and in the "good" range for IIHS rating scheme
FMVSS Pole	Comparable to the baseline Honda Accord	Comparable to baseline Honda Accord and LWV
IIHS Side Impact with Moving Deformable Barrier	Comparable to the baseline Honda Accord	Comparable to baseline Honda Accord and LWV in the IIHS "good' range
IIHS roof crush	Strictly a static test and not a dynamic examination	Comparable to the baseline Honda Accord and LWV in the IIHS "good" range
FMVSS No. 301 Rear Impact Test	Meaningful comparison not possible since no rear impact tests have been run on the baseline Honda Accord	Fuel leakage unlikely because fuel system was not damaged

Figure 5.41 Crash performance comparison of the NanoSteel body structure

Since the thickness of the various body components were changed during the safety optimization, the static stiffness and the modal results had to be rerun to see if the results were still acceptable. The final results are shown in Figures 5.42, 5.43, and 5.44. As suspected, the final results came out to be even better than the post-GENESIS results. The only number in the previous runs which was a bit low was the torsional stiffness, which actually exceeded the Honda Accord number in the final run.

Vehicle	Torsion Stiffness (kN-m/deg)
2011 Honda Accord Test (Target)	12.33
LWV	14.40
NanoSteel Initial Run	10.73
NanoSteel Genesis Optimization	10.49
NanoSteel Final Design	12.98

Figure 5.42 Final torsional stiffness comparison

Vehicle	Bending Stiffness (N/mm)
2011 Honda Accord Test (Target)	8,690
LWV	11,760
NanoSteel Initial Run	8,366
NanoSteel Genesis Optimization	8,510
NanoSteel Final Design	10,641

Figure 5.43 Final bending stiffness comparison

Vehicle	Front-End Lateral Mode (Hz)	Second Order Bending Mode (Hz)	First Order Bending Mode (Hz)	Torsion Mode (Hz)	BIW Mass (kg)
2011 Honda Accord Test Frequency (Target)	35.1	39.3	44.2	50.1	328.0
LWV Modeled Frequency	40.5	40.8	46.3	48.7	255.2
NanoSteel Initial Run Modeled Frequency	34.2	36.6	40.8	45.5	203.0
NanoSteel Genesis Optimized Modeled Frequency	35.7	39.0	43.5	48.5	211.0
NanoSteel Final Design Modeled Frequency	35.7	39.8	44.6	50.4	228.1

Figure 5.44 Final normal mode frequencies results comparison

The weight benefit of the NanoSteel body-in-white was then compared with the baseline Honda (see Figure 5.45). The weight reduction from the baseline now stood at 30% versus 22% for the LWV. To see how the individual steels in the NanoSteel portfolio contributed to the overall weight reduction, Figure 5.46 shows the N1 contribution, Figure 5.47 shows the N2 contribution, and Figure 5.48 shows the N3 contribution. The N1 and N2 material substitutions had fairly large percentage reductions from the LWV, and N3 had a fairly small % reduction. However, many of the parts where N3 was used were hot-stamped boron steel in the LWV, which also has very high strength. However, in production N3 may be used more heavily if it turns out that the price of N3 is significantly below the material price of HSB plus the extra processing costs of the hot stamping.

Vehicle	Body-in-White Mass (kg)	Weight Reduction (%)
2011 Honda Accord	328.0	–
NHTSA LWV	255.2	22%
NanoSteel Body-in-White	228.1	30%

Figure 5.45 NanoSteel study mass savings summary

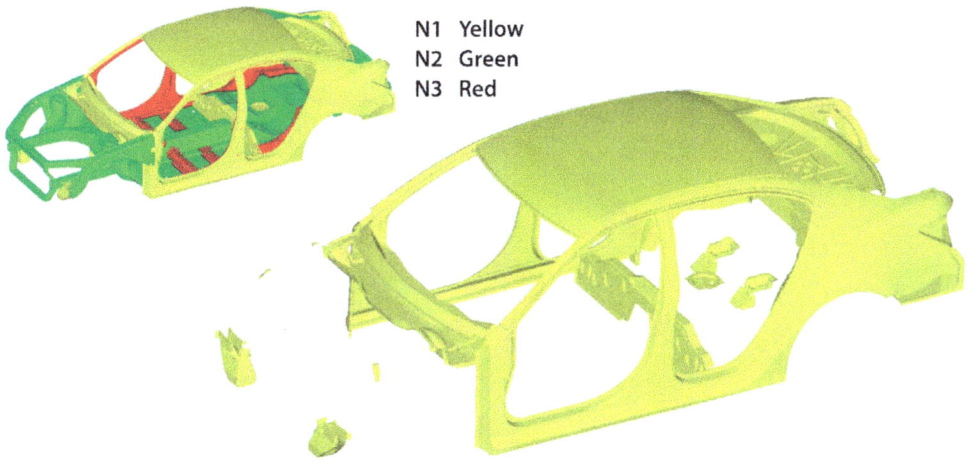

Figure 5.46 N1 grade used for external body panels

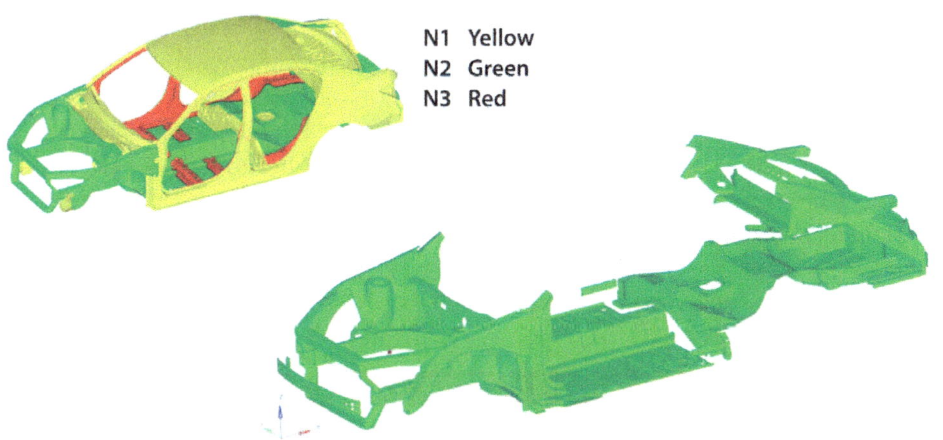

Figure 5.47 N2 grade used for front and rear crash rails and floor structure

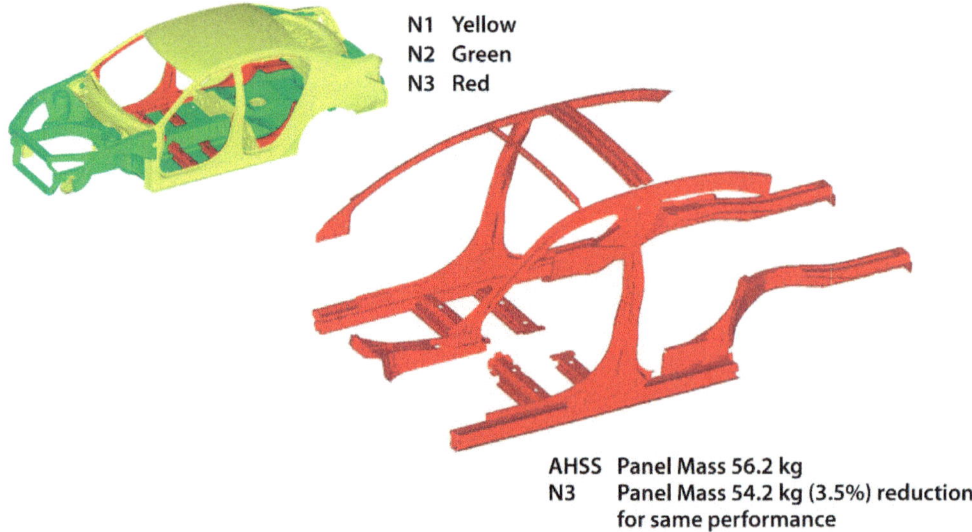

Figure 5.48 N3 grade used for very high strength parts in roof and body side

To help clarify these results, EDAG ran two other conditions, labeled "B" and "C" in the table in Figure 5.49, with the results shown in Figure 5.50. As suspected, replacing only UHSS with N3 and leaving the other LWV steels alone resulted in very little additional weight reduction. Adding N2 and N3 but leaving out N1 resulted in about a 4% additional benefit versus the baseline vehicle, and finally adding N1 contributes another 4%. The N1 and N2 materials contribute approximately the same weight savings in the vehicle. If it is assumed that the LWV architecture and materials become the norm for bodies-in-white in a couple of years, then the aluminum intensive vehicle provides only a 17% weight reduction from the LWV. At the point in which third generation AHSS bodies-in-white become the norm, then aluminum provides about 7% weight reduction from that baseline.

Option	Details
A	LWV - All AHSS (from the NHTSA Study)
B	All of the Ultra High Strength Steels (UHSS) over 1000 MPa, including the hot stamped parts, are replaced with the N3. Other steels remain
C	N2 and N3 are used in place of UHSS, AHSS and conventional steels in the LWV design where applicable, others remain.
D	N1, N2 and N3 grades replace all steels (the option discussed throughout the report)
E	All aluminum (from the NHTSA Study)

Figure 5.49 Light-weighting options reviewed

BIW Material Options	2011 Honda Accord BIW Mass (kg)	BIW Mass (kg)	Mass Saving (kg)	Mass Reduction (%)
A Advanced High-Strength Steel (AHSS)	328	255.2	72.8	- 22.2%
B Option A, adding N3 for all UHSS	328	253.3	74.7	- 22.8%
C Option B, adding N2	328	240.6	87.4	- 26.6%
D All NanoSteel Vehicle	328	228.1	99.9	- 30.5%
E Aluminum Intensive	328	213.0	114.8	- 35.0%

Figure 5.50 Options for Using NanoSteel's sheet steel versus other materials

In conclusion:

1. NanoSteel's automotive sheet has the potential to provide weight reduction beyond that provided by first generation AHSS. This can be achieved without compromising safety standards or NVH, through higher combinations of tensile and elongation properties that enable thinner part gauges.
2. The primary benefit of using NanoSteel's AHSS in the body-in-white is a weight savings of over 30% compared with the baseline 2011 Honda Accord. This weight savings is also 10.5% higher than what was achieved using the current steels in the LWV design.
3. Weight savings in the body structure have total vehicle implications. A lighter vehicle can use a smaller powertrain to deliver the same performance, which means in turn that the supporting structure can be smaller and lighter. Also, the vehicle suspension system can be lighter. These compounding effects on the requirements of the body structure were not taken into account in the EDAG study, but could result in greater weight savings.
4. Another benefit of the use of NanoSteel's AHSS is a reduction in the use of hot-formed parts, which require expensive pre- and post-processing. This could potentially reverse current steel trends in which hot-forming steels are currently the fastest growing AHSS for automotive parts.
5. No redesign of the overall vehicle body or part architecture occurred during this study beyond that which had been derived for the LWV design in the original NHTSA study. In a full redesign, in which architecture was optimized to the NanoSteel steel characteristics, more weight might be removed.
6. For NanoSteel grades N1 and N2, there is high workhardening, which would be experienced upon forming parts. To capture the benefit of this feature, forming analysis would need to be conducted before crash optimization and new element-by-element stress-strain curves would have been derived prior to crash analysis. For the most part, this would have resulted in stronger steel in the final analysis, which could have potentially resulted in more weight savings.

7. Since an aluminum vehicle that was analyzed in the original EDAG NHTSA study only achieved 4.6% reduction from the baseline vehicle incremental to that, which was achieved from the NanoSteel vehicle, the high cost of aluminum will likely make the NanoSteel AHSS more competitive for high-volume mass market vehicle platforms. When combined with other benefits such as the cost of repairs, manufacturers will likely remain in steel for body-in-white structures for some time to come.

5.5 References

5-1. Grabowski, Thomas, 2013, "2014 Chevrolet Silverado/ GMC Sierra Body 1500 Cab Structure Review," presentation to the 2013 Great Designs in Steel Conference, Livonia, Michigan, May 2013, available at www.autosteel.org, accessed in August 2013.

5-2. Matlock, D. K. and Spear, J. G., "Third Generation of AHSS: Microstructure Design Concepts," Proceedings of the International Conference on Microstructure and Texture in Steels and Other Materials, ed. By A. Haldar, S. Suwas, and D. Bhattacharjee, Springer, London, 2009, pp. 185–205.

5-3. Branagan, Daniel, "Overview of a New Category of Third Generation AHSS," presentation to the 2013 Great Designs in Steel Conference, Livonia, Michigan, May 2013, available at www.autosteel.org, accessed in August 2013.

Chapter 6
Conclusions and Recommendations

1. The advanced high-strength steels (AHSS) have typically more strength than the conventional high-strength steels (CHSS) (e.g., HSLA, BH, and so on) and/or have more ductility than the CHSS at the same strength level. Dual-phase (DP), complex phase (CP), transformation induced plasticity (TRIP), and the martensitic (Mart) are all considered AHSS.

2. The primary driver for the use of AHSS is vehicle safety. There has been a steady progression of new National Highway & Traffic Safety Administration (NHTSA) required safety regulations and IIHS nonregulatory test procedures going back to the early 1990s through to the present time. The high strength to yield ratio of DP AHSS provides energy absorption, and the high strength of martensitic AHSS reduces intrusion.

3. Weight reduction is equally responsible for the increased use of AHSS because it directly affects fuel economy. The Environmental Protection Agency (EPA) has set a goal of achieving an average of 54.5 MPG (14.4 km/l) by 2025. AHSS have been demonstrated to provide significant weight reduction in several weight reduction studies cited in this book.

4. Another important driver for increasing fuel economy is the reduction of greenhouse gas emissions (e.g., CO_2). Reducing greenhouse gas emissions using AHSS has been shown to be a better strategy than using alternative materials. AHSS, like all steel, is 100% recyclable; steel remains the most recycled material; steel recycling has a large infrastructure in terms of dismantlers and shredders; and steel autos enjoy a 97% end-of-life recycling rate.

5. AHSS currently have less than a 10% penetration as a measure of all the steel used in the vehicle across the whole automotive industry. However, because of regulatory concerns, the current growth rate of AHSS, and various benchmark studies (e.g., FSV), the penetration by 2025 could be greater than 80%. The highest growth rate currently is in hot-stamped boron body applications. Chassis applications have lagged body applications significantly.

6. The number one impediment for using AHSS is forming problems. Even though AHSS are more formable than equivalent tensile strength CHSS, there are major forming problems because typically one is using higher strength AHSS than the CHSS being replaced. There are a number of well-published countermeasures to offset AHSS forming problems. Using four-piece dies (i.e., using pads) versus three-piece or two-piece will facilitate stamping.

7. The second largest impediment to the use of AHSS is welding, though it is not nearly as large of an impediment as formability. The conversion from AC welding to mid-frequency DC (MFDC) welding will solve many of the AHSS welding problems. Weld cap optimization (e.g., spherical caps) and increasing weld time will also facilitate AHSS welding.
8. Another impediment to using AHSS is the fatigue life of the welds joining the AHSS. The fatigue life of the welds is closer to mild steel because of the annealing, which goes on during the welding process. Using more welds in the case of spot welding or a longer weld line in the case of mig welding will improve this situation. Also, redesign to move the high stress points away from the weld can also be an effective tool.
9. Another potential impediment is the loss of stiffness, which can create noise, vibration, and harshness problems, when the thickness of AHSS panels is reduced to optimize weight. Computer-aided engineering (CAE) thickness optimization tools can minimize this problem. Also, improvements in architecture and joining to improve stiffness retention can be helpful. As far as joining is concerned, using laser welding or structural adhesives can make large improvements in stiffness retention. Again, CAE tools can be used to locate critical seams that can benefit from improved joining. Often only a small percentage of the weld seams (e.g., 25%) need to use structural adhesive or laser welding to get nearly 100% of the benefit.
10. There are several architectural enablers or fabrication methods that can be used synergistically with AHSS to enhance the overall design from a safety and weight minimization perspective. Among these are tube hydroforming, roll forming, Tailor rolled blanks, laser-welded blanks, and hot stamping. Laser-welded blanks have been around for a long time and have been commonly used for door inners and front and rear body rails. The fastest growing enabler is hot stamping using hot-forming boron steel. This overcomes the number one impediment to using AHSS, which is forming difficulties, and can provide extremely strong (1500 MPa tensile strength) parts. However, there is extra processing time and costs associated with this practice.
11. Aluminum is the closest material competitor to steel for automotive applications. For closed section tubes with the same outside diameter there is no weight advantage for a conversion from steel to aluminum for the same bending and torsional stiffness and Euler buckling. For plates with the same bending and torsional stiffness and the same area, there is a 50% weight reduction for a conversion from steel to aluminum. Therefore, space frame architectures, where most of the load is carried by the steel skeleton, will have the least advantage for a conversion from steel to aluminum.
12. There is a metric (λ) that can be used to compute, using a CAE model, the weight advantage of converting a steel structure to aluminum for a given loading condition and responsee. It will vary, depending on the specific structure being analyzed, between 1 and 3. For a structure in which λ is close to 1, when analyzed for a specific loading condition, there will be no weight advantage for a conversion to aluminum. For a structure in which λ is close to 3, there will be a 50% weight reduction with a conversion to aluminum. The more robust the structure from a load bearing perspective, the closer

λ will be to 1. This means that λ can be used to compute the weight efficiency of the structure and avoid costly steel to aluminum conversion.

13. According to a 2012 NHTSA/EDAG study on the 2011 Honda Accord, an analytical conversion of the body-in-white to improved architectures/joining methods and a conversion to the latest level of AHSS yielded a 23% weight reduction. A conversion of the body-in-white to aluminum resulted in a 35% weight reduction from the 2011 Honda Accord. NHTSA/EDAG elected to stay with steel for the body-in-white because of the high cost to get a little weight improvement in aluminum.
14. As part of the NHTSA/EDAG study, a door was analyzed, and it resulted in a 15% weight reduction for a conversion to improved architecture and AHSS, which was consistent with an Auto/Steel Partnership study, which was done earlier. A conversion to aluminum resulted in a 48% weight reduction. Given the 2025 fuel economy improvement goal of 54.5 MPG, closures could be converted to aluminum in the near future. However, if doors were designed to be better integrated with the body-in-white structure, the advantage of a conversion to aluminum could be diminished.
15. As part of the NHTSA/EDAG study and a number of other studies, weight reduction of suspension arms, frames, and wheels should be about equal for conversion to AHSS or aluminum. Of course, the cost advantage of steel will typically win out.
16. Using the cost data from the NHTSA/EDAG study, a high grade of AHSS costs about $0.67/lb, austenitic stainless costs about $2.10/lb, and aluminum sheet costs about $2.14/lb. Press-hardened boron steel with the extra processing expense (over and above conventionally stamped steels) costs about $1.80/lb. It is easy to see why auto manufacturers are not using stainless steel except in exhaust systems, where they are using a cheaper grade of stainless. Also, press-hardened steel (PHS) even with the extra processing costs is considerably cheaper than aluminum. There is a window of opportunity for a new grade of steel (e.g., third generation AHSS) priced somewhere in the $0.67/lb and the $1.84/lb range if it can be shown to be as strong as 1500 MPa tensile strength and as formable as a lower grade of AHSS. If this steel came to fruition, use of PHS may level off.
17. Steel companies, universities, national labs, and independent labs are working on a new steel, which is super strong, super ductile, and moderate in cost, and is called third generation AHSS. This is because second generation AHSS, which is also similar in strength and ductility, never came to automotive production because of the high cost (e.g., austenitic stainless steel) and production issues. Several strategies for making this steel are mentioned in this book, with some detail supplied for NanoSteel's new steel.
18. A recent EDAG study using NanoSteel's AHSS properties showed that the 2011 Honda Accord's body-in-white could be reduced in weight by 30.5%. This compares to the previously mentioned NHTSA/EDAG study in which the body was reduced by 35% using aluminum.
19. The calculated gap between steel and aluminum weight reduction has diminished tremendously since the Program for a New Generation of Vehicles of the early 1990s. This is because aluminum materials have remained essentially the same, and aluminum

is a more architecturally robust material because of the thickness advantage over steel. This architectural advantage for aluminum means that there is less opportunity for aluminum to provide further weight improvements for improved architecture. Meanwhile, automotive steel architecture and improved steel grades are causing the difference in weight between steel and aluminum construction to diminish, and it probably will continue to diminish in the future.

Index

About the Author

Paul Geck was the senior staff advanced high-strength steel technical specialist at Ford Motor Company before he retired. At the time of his retirement, he was also the chairman of the Auto Steel Partnership. Mr. Geck has a track record in leading major cross-functional, engineering research projects oriented toward developing new engineering technologies and for developing implementation strategies for those technologies. For instance, he led the Improved Materials & Powertrain Architectures for 21st Century Trucks (IMPACT) Project. This was a large-scale, cross-industry/Department of Defense project, which focused on pick-up truck weight reduction using new steel technology. Currently, Mr. Geck teaches the Advanced High-Strength Steel Seminar for SAE International; and since retirement from Ford he has worked as a consultant to the steel and automotive industries as an advanced high-strength steel expert. He has two masters degrees in engineering and an MBA from the University of Michigan and has numerous publications in the fields of computer-aided engineering, noise and vibration engineering, steel technology, and automotive weight reduction. Mr. Geck was awarded an Industry Leadership Award by the American Iron & Steel Institute in 2006 and was awarded a Long-Term Contribution Award by the Auto/Steel Partnership in 2007. He earned an SAE-Automotive Resources Institute Registered Consultant Designation in 2008.